Building Design and Construction Systems (BDCS)

ARE Mock Exam (Architect Registration Exam)

ARE Overview, Exam Prep Tips,
Multiple-Choice Questions and Graphic Vignettes,
Solutions and Explanations

Gang Chen

ArchiteG®, Inc.
Irvine, California

Building Design and Construction Systems (BDCS) ARE Mock Exam (Architect Registration Exam): ARE Overview, Exam Prep Tips, Multiple-Choice Questions and Graphic Vignettes, Solutions and Explanations

ArchiteG®, Inc.
http://www.ArchiteG.com

ISBN: 978-1-61265-002-9

PRINTED IN THE UNITED STATES OF AMERICA

Dedication

To my parents, Zhuixian and Yugen,
my wife, Xiaojie, and my daughters,
Alice, Angela, Amy, and Athena.

Disclaimer

How to Use This Book

We suggest you read *Building Design and Construction Systems (BDCS) ARE Mock Exam (Architect Registration Exam)* at least three times:

Read once, and cover Chapter One, Two and Appendixes and the related FREE PDF files and other resources. Highlight the information with which you are not familiar.

Read a second time, this time focusing on the highlighted information to memorize. You can repeat this process as many times as you want until you have mastered the content of the book. Pay special attention to the materials listed in Chapter Two, Section B, **The most important documents/publications for BDCS division of the ARE exam.**

After reviewing these materials, you can do the mock exam, and then check your answers against the answers and explanations in the back; including explanations for the questions you answered correctly. You may have answered some questions correctly, but for the wrong reason. Highlight the information you are not familiar with.

Like the real exam, the mock exam includes all three types of questions: Select the correct answer, check all that apply, and fill in the blank.

Review your highlighted information, and do the mock exam again. Try to answer 100% of the questions correctly this time. Repeat the process until you can answer all of the questions correctly.

Do the mock exam about two weeks before the real exam, but at least 3 days before the real exam. You should NOT wait until the night before the real exam, and then do the mock exam: If you do not do well, you will go into panic mode and you will NOT have enough time to review your weakness.

Read for the third time the night before the real exam. Review ONLY the information you highlighted, especially the questions you did not answer correctly when you did the mock exam for the first time.

One important tip for passing the graphic vignette section of the ARE BDCS division is to become VERY familiar with the commands of the NCARB software. Many people fail the exam simply because they are NOT familiar with the NCARB software and cannot finish the graphic vignette section within the exam's time limit.

For the graphic vignettes, we include step-by-step solutions, using NCARB Practice Program software, with many screen-shots so that you can use this book to become familiar with the commands of the NCARB software, even when you do NOT have a computer in front of you.

This book is very light and you can carry it around easily. These two features will allow you to review the graphic vignette section whenever you have a few minutes.

All commands are described in an **abbreviated manner**. For example, **Draw > Stairs** means go to the menu on the left hand side of your computer screen, click **Draw,** and then click **Stairs** to draw the **Stair.** This is typical for ALL commands throughout the book.

The Table of Contents is very detailed, so you can locate information quickly. If you are on a tight schedule, you can forgo reading the book linearly and jump to the sections you need.

All our books including "ARE Mock Exams Series" and "LEED Exam Guides Series" are available at
GreenExamEducation.com

Check out FREE tips and info at **GeeForums.com**, you can post your questions or vignettes for other users' review and responses.

Table of Contents

Chapter One   **Overview of Architect Registration Exam (ARE)**

   1. Important links to the FREE and official NCARB documents
   2. A detailed list and brief description of the FREE PDF files that you can download from NCARB
   - ARE Guidelines
   - NCARB Education Guidelines
   - Intern Development Program Guidelines
   - The IDP Supervisor Guidelines
   - Handbook for Interns and Architects
   - Official exam guide, <u>references index</u>, and practice program (NCARB software) for each ARE division
   - The Burning Question: Why Do We Need ARE Anyway?
   - Defining Your Moral Compass
   - Rules of Conduct

   1. What is IDP?
   2. Who qualifies as an intern?

   1. How to qualify for the ARE?
   2. How to qualify for an architect license?
   3. What is the purpose of the ARE?
   4. What is NCARB's rolling clock?
   5. How to register for an ARE exam?
   6. How early do I need to arrive at the test center?

7. Exam Format & Time
   - Programming, Planning & Practice
   - Site Planning & Design
   - Building Design & Construction Systems
   - Schematic Design
   - Structural Systems
   - Building Systems
   - Construction Documents and Service
8. How are ARE scores reported?
9. Are there a fixed percentage of candidates who pass the ARE exams?
10. When can I retake a failed ARE division?
11. How much time do I need to prepare for each ARE division?
12. Which ARE division should I start first?
13. ARE exam prep and test-taking tips
14. English system (English or inch-pound units) vs. metric system (SI units)
15. Codes and standards used in this book

Chapter Two   Building Design and Construction Systems (BDCS) Division

Back Page Promotion

Index

Chapter One

Overview of Architect Registration Exam (ARE)

A. First Thing First: Go to the website of your architect registration board and read all the requirements for obtaining an architect license in your jurisdiction.
See following link:
http://www.ncarb.org/Getting-an-Initial-License/Registration-Board-Requirements.aspx

B. Download and review the latest ARE documents from NCARB website

1. Important links to the FREE and official NCARB documents
The current version of Architect Registration Exam includes seven divisions:

- Programming, Planning & Practice
- Site Planning & Design
- Building Design & Construction Systems
- Schematic Design
- Structural Systems
- Building Systems
- Construction Documents and Service

Note: Starting July 2010, 2007 AIA Documents apply to ARE Exams.

Six ARE divisions have a multiple-choice section and a graphic vignette section. Schematic Design division has NO multiple-choice section, but two graphic vignette sections.

For the vignette section, you need to create the following graphic vignette(s) based on the ARE division you are taking:

Programming, Planning & Practice
Site Zoning

Site Planning & Design
Site Grading
Site Design

Building Design & Construction Systems
Accessibility/Ramp
Stair Design
Roof Plan

Schematic Design
Interior Layout
Building Layout

Structural Systems
Structural Layout

Building Systems
Mechanical & Electrical Plan

Construction Documents & Services
Building Section

There is a tremendous amount of valuable information covering every step of becoming an architect available free of charge at NCARB website:
http://www.ncarb.org/

For example, you can find the education guide regarding professional architectural degree programs accredited by the National Architectural Accrediting Board (NAAB), NCARB's Intern Development Program (IDP) guides, initial license, certification and reciprocity, continuing education, etc. These documents explain how you can become qualified to take the Architect Registration Exam.

I find the official ARE Guidelines, exam guide and practice program for each of the ARE divisions extremely valuable. See the following link:
http://www.ncarb.org/ARE/Preparing-for-the-ARE.aspx

You should definitely start by studying the official exam guide and practice program for the ARE division you are taking.

2. **A detailed list and brief description of the FREE PDF files that you can download from NCARB**
The following is a detailed list of the FREE PDF files that you can download from NCARB. We list them in an order based on their importance:

- **ARE Guidelines**: Very important, includes extremely valuable information on the ARE overview, six steps to complete ARE, multiple-choice section, graphic vignette section, exam format, scheduling, sample exam computer screens, links to other FREE NCARB PDF files, practice software for graphic vignettes, etc. You need to read this document at least twice.

- **NCARB Education Guidelines** (Skimming through it should be adequate)

- **Intern Development Program Guidelines**: Important information on IDP overview, IDP steps, IDP reporting, IDP basics, work settings, training requirements, supplementary education (core), supplementary education (elective), core competences, next steps, and appendices. Most of NCARB's 54-member boards have adopted the IDP as a prerequisite for initial architect licensure. That is the reason you should care about it. IDP costs $350 for the first three years, and then $75 annually. The fees are subject to change, and you need to check the NCARB website for the latest information. You need to report your IDP experience <u>no longer than every six months</u> and within two months of completion of each reporting period (the **Six-Month Rule**). You need to read this document <u>at least twice</u>. It has a lot of valuable information.

- **The IDP Supervisor Guidelines** (Skimming through it should be adequate. You should also forward a copy of this PDF file to your IDP supervisor.)

- **Handbook for Interns and Architects** (Skimming through it should be adequate)

- **Official exam guide, <u>references index</u>, and practice program (NCARB software) for each ARE division**
  Specific information for ARE divisions (You just need to focus on the documents related to the ARE divisions you are currently taking and read them at least twice. Make sure you install the practice program and become very familiar with it. The real exam is VERY similar to the practice program):

  a. **Programming, Planning & Practice (PPP)**: Official exam guide and practice program for PPP division
  b. **Site Planning & Design (SPD)**: Official exam guide and practice program (computer software) for SPD division
  c. **Building Design & Construction Systems (BDCS)**: Official exam guide and practice program for BDCS division
  d. **Schematic Design (SD)**: Official exam guide and practice program for SD division
  e. **Structural Systems (SS)**: Official exam guide, <u>references index</u>, and practice program for SS division
  f. **Building Systems (BS)**: Official exam guide and practice program for BS division
  g. **Construction Documents and Service (CDS)**: Official exam guide and practice program for CDS division

- **The Burning Question: Why Do We Need ARE Anyway?** (Skimming through it should be adequate)

- **Defining Your Moral Compass** (Skimming through it should be adequate)

- **Rules of Conduct** (Skimming through it should be adequate). Available as a FREE PDF file at:
  http://www.ncarb.org/

C. The Intern Development Program (IDP)

1. What is IDP?

It is a comprehensive training program jointly developed by the National Council of Architectural Registration Boards (NCARB) and the American Institute of Architects (AIA) to ensure the interns obtain the necessary skills and knowledge to practice architecture <u>independently</u>.

2. Who qualifies as an intern?

Per NCARB, if an individual meets one of the following criteria, s/he qualifies as an intern:

a. Graduates from NAAB-accredited programs
b. Architecture students who acquire acceptable training prior to graduation
c. Other qualified individuals identified by a registration board

D. Overview of Architect Registration Exam (ARE)

1. How to qualify for the ARE?

A candidate needs to qualify for the ARE via one of NCARB's member registration boards, or one of the Canadian provincial architectural associations.

Check with your Board of Architecture for specific requirements.

For example, in California, a candidate must provide verification of a minimum of <u>five</u> years of education and/or architectural work experience to qualify for the ARE.

Candidates can satisfy the five-year requirement in a variety of ways:

- Provide verification of a professional degree in architecture through a program that is accredited by NAAB or CACB.

  OR
- Provide verification of at least five years of educational equivalents.

  OR
- Provide proof of work experience under the direct supervision of a licensed architect

2. **How to qualify for an architect license?**

Again, each jurisdiction has its own requirements. It typically requires a combination of about <u>eight</u> years of education and experience, as wells as passing ARE exams. See following link:

http://www.ncarb.org/Reg-Board-Requirements

For example, to become a licensed architect in California, you need:
- Eight years of post-secondary education and/or work experience as evaluated by the Board (including at least one year of work experience under the direct supervision of an architect licensed in a U.S. jurisdiction or two years of work experience under the direct supervision of an architect registered in a Canadian province)
- Completion of the Comprehensive Intern Development Program (CIDP) and the Intern Development Program (IDP)
- Successful completion of the Architect Registration Examination (ARE)
- Successful completion of the California Supplemental Examination (CSE)

California does NOT require an accredited degree in architecture for examination and licensure. However, many other states require an accredited degree for licensure.

3. **What is the purpose of the ARE?**

The purpose of ARE is NOT to test a candidate's competency on every aspect of architectural practice. Its purpose is to test a candidate's competency on providing professional services to protect the <u>health, safety, and welfare</u> of the public. It tests candidates on the <u>fundamental</u> knowledge of pre-design, site design, building design, building systems, and construction documents and services.

ARE tests a candidate's competency as a "specialist" on architectural subjects. It also tests her abilities as a "generalist" to coordinate other consultants' works.

You can download the exam content and references for each of the ARE divisions at the following link:

http://www.ncarb.org/are/40/StudyAids.html

4. **What is NCARB's rolling clock?**
   a. Starting on January 1, 2006, a candidate MUST pass ALL ARE sections within 5 years. A passing score for an ARE division is only valid for 5 years, and a candidate has to retake this division if he has NOT passed all divisions within the 5-year period.

   b. Starting on January 1, 2011, a candidate who is authorized to take ARE exams MUST take at least one division of the ARE exams within 5 years of the authorization. Otherwise, the candidate MUST apply for the authorization to take ARE exams from an NCARB member board again.

These rules are created by the **NCARB's rolling clock** resolution passed by NCARB council in 2004 NCARB Annual Meeting.

5. **How to register for an ARE exam?**
   Go to the following website and register:
   http://www.prometric.com/NCARB/default.htm

6. **How early do I need to arrive at the test center?**
   At least 30 minutes BEFORE your scheduled test time, OR you may lose you exam fee.

7. **Exam Format & Time**
   All ARE divisions are administered and graded by computer. Their detailed exam format and testing time are as follows:

1) **Programming, Planning & Practice (PPP)**
   The **Programming, Planning & Practice** division of the Architect Registration Exam (ARE) includes a multiple-choice (MC) section and a graphic vignette section, and lasts a total of 4 hours. It includes:

Introduction Time:	15 minutes	
MC Testing Time:	**2 hours**	**85 items**
Scheduled Break:	15 minutes	
Introduction Time:	15 minutes	
Graphic Testing Time:	**1 hour**	**Site Zoning (1 vignette)**
Exit Questionnaire:	15 minutes	
Total Time	**4 hours**	

2) **Site Planning & Design (SPD)**

Introduction Time:	15 minutes	
MC Testing Time:	**1.5 hours**	**65 items**
Scheduled Break:	15 minutes	
Introduction Time:	15 minutes	
2 Graphic Vignettes:	**2 hours**	**Site Grading, Site Design**
Exit Questionnaire:	15 minutes	
Total Time	**4.5 hours**	

3) **Building Design & Construction Systems (BDCS)**

Introduction Time:	15 minutes	
MC Testing Time:	**1.75 hours**	**85 items**
Scheduled Break:	15 minutes	
Introduction Time:	15 minutes	
3 Graphic Vignettes:	**2.75 hours**	**Accessibility/Ramp, Stair Design, Roof Plan**
Exit Questionnaire:	15 minutes	
Total Time	**5.5 hours**	

4) **Schematic Design (SD)**

Introduction Time:	15 minutes	
Graphic Testing Time:	**1 hour**	**Interior Layout (1 vignette)**
Scheduled Break:	15 minutes	
Introduction Time:	15 minutes	
Graphic Testing Time:	**4 hours**	**Building Layout (1 vignette)**
Exit Questionnaire:	15 minutes	
Total Time	**6 hours**	

5) **Structural Systems (SS)**

Introduction Time:	15 minutes	
MC Testing Time:	**3.5 hours**	**125 items**
Scheduled Break:	15 minutes	
Introduction Time:	15 minutes	
Graphic Testing Time:	**1 hour**	**Structural Layout (1 vignette)**
Exit Questionnaire:	15 minutes	
Total Time	**5.5 hours**	

6) **Building Systems (BS)**

Introduction Time:	15 minutes	
MC Testing Time:	**2 hours**	**95 items**
Scheduled Break:	15 minutes	
Introduction Time:	15 minutes	
Graphic Testing Time:	**1 hour**	**Mechanical & Electrical Plan (1 vignette)**
Exit Questionnaire:	15 minutes	
Total Time	**4 hours**	

7) **Construction Documents and Service (CDS)**

Introduction Time:	15 minutes	
MC Testing Time:	**2 hours**	**100 items**
Scheduled Break:	15 minutes	
Introduction Time:	15 minutes	
Graphic Testing Time:	**1 hour**	**Building Section (1 vignette)**
Exit Questionnaire:	15 minutes	
Total Time	**4 hours**	

8. **How are ARE scores reported?**

All ARE scores are reported as Pass or Fail. ARE scores are processed within 4 to 6 weeks, and then sent to your Board of Architecture. Your board then does additional processing and forwards the scores to you.

9. Are there a fixed percentage of candidates who pass the ARE exams?
No, there is NOT a fixed passing or failing percentage. If you meet the minimum competency required to practice as an architect, you pass. The passing scores are the same for all Boards of Architecture.

10. When can I retake a failed ARE division?
You can only take the same ARE division once within a 6-month period.

11. How much time do I need to prepare for each ARE division?
You need about 40 hours to prepare for each ARE division. You need to set a realistic study schedule and stick with it. Make sure you allow time for personal and recreational commitments. If you are working full time, my suggestion is that you allow no less than 2 weeks but NOT more than 2 months to prepare for each ARE division. You should NOT drag out the exam prep process too long and risk losing your momentum.

12. Which ARE division should I start first?
It is a matter of personal preference, and you should make the final decision.

Some people like to start with the easier divisions and pass them first. This way, they build more confidence as they study and pass each division.

Some people like to start with the more difficult divisions and pass them first. This way, if they fail a difficult, as they study the other divisions, the clock start to clicks, and they can reschedule the failed division six months later.

Programming, Planning & Practice (PPP) and Building Design & Construction Systems (BDCS) divisions often include some content from the Construction Documents and Service (CDS) division. It may be a good idea to start with CDS and then schedule the exams for PPP and BDCS soon after.

13. ARE exam prep and test-taking tips
You can start with Construction Documents and Service (CDS) and Structural Systems (SS) first because both divisions gave a limited scope, but you may want to study building regulations and architectural history (especially famous architects and buildings that set the trends at critical turning points) before you take other divisions.

Do mock exams and practice questions and vignettes, including those provided by NCARB's practice program and this book, to hone your skills.

Form study groups and learn the exam experience of other ARE candidates. The forum at our website is a helpful resource. See the following link:
http://GreenExamEducation.com/

Take the ARE exams as soon as you become eligible, since you probably still remember portions of what you learned in architectural schools, especially structural and architectural history. Do not make excuses for yourself and put off the exams.

The following test-taking tips may help you:
- Pace yourself properly: You should spend about one minute for each Multiple-Choice (MC) question, except for the SS division. You should spend about one and a half minutes for each Multiple-Choice (MC) question for the SS division.
- Read the questions carefully and pay attention to words like *best, could, not, always, never, seldom, may, false, except,* etc.
- For questions that you are not sure of, eliminate the obvious wrong answer and then make an educated guess. If you do NOT answer the question, you automatically lose the point. If you guess, you have a chance to gain the point.
- If you have no idea what the correct answer is and cannot eliminate any obvious wrong answer, then do not waste too much time on the question; just pick a guess answer. The key is, try to use the same guess answer for all of the questions for which you have no ideas at all. For example, if you choose "d" as the guess answer, then you should be consistent and use "d" as the guess answer for all the questions for which you have no ideas at all. That way, you likely have a better chance at guessing more answers that are correct.
- Mark the difficult questions, answer them, and come back to review them AFTER you finish all MC questions. If you are still not sure, go with your first choice. Your first choice is often the best choice.
- You really need to spend time practicing to become VERY familiar with NCARB's graphic software and know every command well. This is because ARE graphic vignette is a timed test, and you do NOT have time to think about how to use the NCARB's graphic software during the test. Otherwise, you will NOT be able to finish your solution to the vignette on time.
- ARE exams test a candidate's competency on providing professional services to protect the <u>health, safety, and welfare</u> of the public. Do NOT waste your time on aesthetic or other design elements not required by the program.

ARE exams are difficult, but if you study hard and prepare well, combined with your experience, IDP training, and/or college education, you should be able to pass all divisions and eventually be able to call yourself an architect.

14. English system (English or inch-pound units) vs. metric system (SI units)
This book is based on the English system or English units; the equivalent value in metric system or SI units follows in parentheses. All SI dimensions are in millimeters unless noted otherwise. The English or inch-pound units are based on the module used in the U.S. The SI units noted are simple conversions from the English units for information only and are not necessarily according to a metric module.

15. Codes and standards used in this book
We use the following codes and standards used in this book:
American Institute of Architects, Contract Documents, Washington, DC.

Canadian Construction Documents Committee, CCDC Standard Documents, 2006, Ottawa.

Chapter Two

Building Design and Construction Systems (BDCS) Division

A. General Information

1. Exam Content

An architect should have the skills and knowledge of building design and construction, including: economic, social and environmental issues, practice and project management.

The exam content for the BDCS division of the ARE includes:
1) **Principles**: Selection of Systems, Materials, and Methods, Historic Precedent, Human Behavior, and Design Theory
2) **Environmental Issues**: Sustainable Design, including Hazardous Material Mitigation, Thermal and Moisture Protection, and Adaptive Re-Use
3) **Codes & Regulations**: Zoning, Specialty and Building Codes, and Other Regulatory Requirements
4) **Materials & Technology**: Selection of Systems, Materials, and Methods, including:
   - Masonry
   - Metals
   - Wood
   - Concrete
   - Other
   - Specialties
5) **Project & Practice Management**: Cost, Scheduling, Construction Sequencing, and Risk Management

For the graphic vignettes, you will be required to design:
1) **Accessibility/Ramp:** Design a stairway and ramp connecting two levels, which abides by the code and accessibility requirements
2) **Stair Design:** Design a stairway connecting multiple levels, which abides by the code and accessibility requirements
3) **Roof Plan:** Design a sloped roof for draining the rainwater, locate equipment and accessories

2. Official exam guide and practice program for BDCS division

You need to read the official exam guide for the BDCS division at least twice. Make sure you install the BDCS division practice program and become very familiar with it. The real exam is VERY similar to the practice program.

You can download the official exam guide and practice program for BDCS division at the following link:
http://www.ncarb.org/en/ARE/Preparing-for-the-ARE.aspx

B. The most important documents/publications for BDCS division of the ARE exam

BDCS is one of those ARE divisions, which is very hard to prepare for by simply reading a finite number of books within a short amount of time. This is because many of the questions in BDCS division are based on your work experience, and it may include questions from other ARE divisions.

We shall help you to alleviate this problem by bringing your attention to some most common issues in architectural practice that related to BDCS division:

BDCS division may include questions from a broad range, but the questions **have to** be issues an **average** architect would encounter during normal architectural practice. NCARB can include some, but not too many, obscure issues. Otherwise, the ARE tests will NOT be **legally defensible**.

Based on our research, the most important documents/publications for BDCS division of the ARE exam are:

1. **Official NCARB list of references for the BDCS division:**
   You can download the NCARB list of references for the BDCS division at the following link:
   http://www.ncarb.org/en/ARE/Preparing-for-the-ARE.aspx

2. ***Building Construction Illustrated* (BCI)**
   Ching, Francis. *Building Construction Illustrated.* Wiley, 2008.
   The illustrations are great! Ching has a great talent for simplifying complicated issues and makes them very easy to understand. You need to at least look through the book and understand the basic components of a building.

   There is a good chance you will get 3 or 4 questions right out of this book in the real BDCS exam.

3. **Historic Preservation documents**
   Based on the readers' feedback, there are quite a few questions regarding historic preservation in the BDCS division exam.

   You should **read** the following documents **at least twice** to become familiar with them. Pay special attention to information in italic fonts and in the shaded area in the PDF file:
   - The two FREE PDF files for *The Secretary of the Interior's Standards for the Treatment of Historic Properties with Guidelines for Preserving, Rehabilitating Restoring & Reconstructing Historic Buildings* and *The Secretary of the Interior's Standards for Rehabilitation & Illustrated Guidelines for Rehabilitating Historic Buildings- Standards* at the following links:

http://www.ironwarrior.org/ARE/Historic_Preservation/
http://www.nps.gov/hps/tps/tax/rhb/masonry01.htm

You should **look through** the following document and become familiar with it:
- *National Historic Preservation Act (NHPA)* at the following link:
  http://www.gsa.gov/portal/content/104441

4. **Access Board,** *ADAAG Manual: A Guide to the American with Disabilities Accessibility Guidelines.* East Providence, RI: BNI Building News. ADA Standards for Accessible Design are available as FREE PDF files at

   http://www.ada.gov/

   AND
   http://www.access-board.gov/adaag/html/figures/index.html

5. **American Institute of Architects (AIA) Documents**
   Reading the summary of the AIA Documents is NOT adequate. You need to read the complete text. Fortunately, you do NOT have to read all the AIA documents.

   Two possible solutions:
   a. Buy ONLY the AIA documents you need from your local AIA office. The following AIA documents are important (especially the AIA documents in **bold** font, read them at least three times. You may have **many** real ARE BDCS division questions based on the following AIA documents listed in **bold** font):

   - **A101–2007, Standard Form of Agreement Between Owner and Contractor where the basis of payment is a Stipulated Sum (CCDC Document 2)**
   - **A201–2007, General Conditions of the Contract for Construction**
   - A503–2007, Guide for Supplementary Conditions
   - A701–1997, Instructions to Bidders (CCDC Document 23)
   - **B101–2007 (Former B141–1997), Standard Form of Agreement Between Owner and Architect (RAIC Document 6)**

     You can find FREE sample forms and commentaries for AIA documents A201 & B101 at the following link:
     http://www.aia.org/contractdocs/aiab081438

   - **C401–2007 (Former C141–1997), Standard Form of Agreement Between Architect and Consultant**

     There are some <u>major changes</u> to the older version of sample C141–1997.

     A FREE older version of sample C141–1997 is available at the following link:
     https://app.ncarb.org/are/StudyAids/_C141.pdf

- G701–2001, Change Order
- G702–1992, Application and Certificate for Payment
- G704–2000, Certificate of Substantial Completion

AIA updates AIA documents roughly every 10 years. Please note that AIA does NOT update <u>all</u> AIA documents at the same time. For example, A701–1997, Instructions to Bidders (CCDC Document 23) is still the most current form.

See the AIA documents price list at the following link for the latest edition of AIA documents:
http://www.aia.org/aiaucmp/groups/aia/documents/pdf/aias076346.pdf

   b. Buy AHPP, and it has a CD that include the sample AIA documents. AHPP itself is also a very important publication for BDCS division. See detailed information at following item.

6. *The Architect's Handbook of Professional Practice* (AHPP)
Demkin, Joseph A., AIA, Executive Editor. *The Architect's Handbook of Professional Practice* (AHPP). The American Institute of Architects & Wiley, latest edition.
This comprehensive book covers all aspects of architectural practice, including two CDs containing the sample AIA contract documents. You may have a few real ARE BDCS division questions, which will be based on this publication. You need to read through this book a few times and know some of the basic architectural practice elements, such as: firm legal structures, marketing and outreach, the delivery methods and compensation (design-bid-build, construction management, and design-build), contracts and agreements, and managing disputes (mediation, arbitration and litigation).

7. *International Building Code* (IBC)
International Code Council, Inc. (ICC, 2006), *International Building Code* (IBC).
You may have a few real ARE BDCS division questions based on this publication. You need to become familiar with some of the more commonly used code sections, such as: allowable areas and allowable areas increase, unlimited areas, egresses, width and numbers of exits required, minimum exit passage width, occupancy groups and related exit occupancy load factors, types of construction, minimum number of required plumbing fixtures required, etc.

See the following link for some FREE IBC code sections citations:
http://publicecodes.citation.com/icod/ibc/2006f2/index.htm?bu=IC-P-2006-000001&bu2=IC-P-2006-000019

8. *Architectural Graphic Standards*
Ramsey, Charles George, and John Ray Hoke Jr. *Architectural Graphic Standards*.
The American Institute of Architects & Wiley, latest edition.
There may be a few questions asking you what some of the basic graphic symbols mean. This is a good book to skim through.

9. **Construction Specifications Institute (CSI) MasterFormat & *Building Construction***
You need to become familiar with the new 6-digit CSI Construction Specifications Institute (CSI) MasterFormat. You may have a few real ARE BDCS division questions based on this publication. You need to know what items/materials belong to which CSI MasterFormat specification section, and remember some major sections names and numbers, such as Division 9 is Finishes, and Division 5 is Metal, etc. My other book, *Building Construction*, has detailed discussions on CSI MasterFormat specification sections.

See the Appendix of this book for other official reference materials suggested by NCARB.

C. Overall strategies and tips for graphic vignettes

1. Overall strategies
To most candidates, the Multiple Choice (MC) portion of an ARE division is harder than the graphic vignettes. Some of the MC questions are based on experience and you do NOT have a set of fixed study materials for them. You WILL make some mistakes on the MC questions no matter how hard you study.

On the other hand, the graphic vignettes are relatively easier, and you do have a good way to prepare for them. You should really take the time to study and practice the NCARB graphic software well, and try to **nail all the graphic vignettes** perfectly. This way, you will have a much better chance to pass even if you answer some MC questions incorrectly.

*Tips: Most people do poorly on the MC portion of the BDCS division, especially those who do NOT have a lot of working experience, but not too many people fail because of the MC portion. Most people fail the ENTIRE BDCS section because they have made **one** fatal mistake on the graphic vignettes section of the BDCS division, such as not leaving enough room for the clerestory window. So, practice the NCARB BDCS practice program graphic software and make sure you absolute NAIL the vignettes section. This is a key for you to pass.*

The official NCARB BDCS exam guide gives a passing solution and a failing solution to each of the sample vignettes, but it does NOT show you the step-by-step details, and it is NOT tied to official NCARB BDCS practice program.

I am going to fill in the blanks here. I am going to offer you step-by-step instructions, and tie them to the commands of the official NCARB BDCS practice program (graphic software).

You really need to spend time practicing to become VERY familiar with NCARB's graphic software. This is because ARE graphic vignette is a timed test, and you do NOT have the time to think about how to use the NCARB's graphic software during the test. Otherwise, you will NOT be able to finish your solution to the vignette on time.

The following solution is based on the official NCARB BDCS practice program for **ARE 4.0**. Future versions of ARE may have some minor changes, but the principles and fundamental elements should be the same. The official NCARB BDCS practice program has very few changes since its introduction and the earlier versions are VERY similar to, or probably exactly the same as, the current ARE 4.0. The actual graphic vignette of the BDCS division should be VERY, VERY similar to the practice graphic vignette in the NCARB BDCS practice program.

2. Tips:
1) You need to install the NCARB BDCS practice program, and become familiar with it. I am NOT going to repeat the vignette description and requirements here since they are already available in the NCARB practice program.

   See the following link for a FREE download of the NCARB practice program:
   http://www.ncarb.org/ARE/Preparing-for-the-ARE.aspx

2) Review the general test directions, vignette directions, program, and tips carefully.
3) Press the space bar to go to the work screen.

D. Accessibility/Ramp Vignette:

1. Major criteria, overall strategy, and tips for accessibility/ramp vignette:
1) Set the minimum dimensions for the **landing** and **width of the ramp** to be **5'-0" (1524)** since the ramp for this vignette typically HAVE to change direction. Set the **width of the stair** to 5'-0" (1524) to match that of the ramp.
2) No portion of the ramp or stair may encroach on the existing upper level. You **need a landing at the top** of your stair and ramp because the top landing is a part of the ramp or stair.
3) Make sure you show the **proper elevations** for the landings.
4) Do NOT forget to extend the ramp non-continuous handrails horizontally at least **12 inches** beyond the **top and bottom** of the ramp run.
5) Do NOT forget to extend the stair non-continuous handrails horizontally at least **12 inches** beyond the **top and bottom** risers.
6) Exit doors shall swing in the direction of egress travel.
7) Maintain enough clearance for both the push side and the pull side of the egress door.

2. Step-by-step solution for the official NCARB BDCS practice program accessibility/ramp vignette:
1) Setting the **length and width of the ramp** as well as the **minimum landing dimension:** Since the elevation difference between the lobby and the upper level is 30" (762), at 1:12 slope, the total length of the ramp is 30' (9144).

Use **Sketch** > **Rectangle** to draw a rectangle covering the entire lobby (Figure 2.1). On the lower left hand corner of the screen, you should see that the dimensions of the rectangle are X: 24'-0" (7315), Y: 24'-0" (7315). This means the lobby size is 24'-0" x 24'-0" (7315 x 7315), and you HAVE to **change the ramp direction** for this vignette since the total length of the ramp is 30' (9144) and larger than 24'-0" (7315). We have to set the minimum landing dimension to **5'-0"** (1524) since the ramp changes direction. For convenience, we set the ramp width to **5'-0"** (1524) also.

2) Click on Cursor to change the cursor to the full-screen cursor; Use **Draw** > **Landing** to draw a **landing #1** covering the entire width of the exit corridor of the upper level, and landing width is 5'-0" (1524, Figure 2.2). You need to click three times: the **first click** set the upper right hand corner of the landing at the exit corridor wall corner; **second click** set the lower right hand corner of the landing at the exit corridor wall corner; **third click** set the horizontal dimension of the landing. On the lower left hand corner of the screen, you can see the dimensions of the rectangle is X: 5'-0" (1524), Y: 6'-0" (1828).

3) Click on **Set Elevation** to bring up the elevation dialogue box (Figure 2.3), use the up arrow to set the **landing #1** elevation to 30" to match the elevation of the upper exit corridor.

4) Use **Draw** > **Landing** to draw a 5'-0" (1524) x 5'-0" (1524) **landing #2** at the lower right-hand corner of the lobby (Figure 2.4).

5) Use **Draw** > **Ramp** to draw a 4'-0" x 5'-0" (1219 x 1524) ramp #1 connecting **landing #1** and **landing #2** (Figure 2.5). Pay attention to the **direction of travel** for the ramp. Since the length of ramp #1 is 4'-0" (1219), the elevation of the landing #2 should be 4" (102) lower than landing #1 = 30"- 4" = 26" (762-102 = 660).

6) Click on **Set Elevation** to bring up the elevation dialogue box, use the up arrow to set the **landing #2** elevation to 26" (660, Figure 2.6).

7) Use **Draw** > **Landing** to draw a 5'-0" x 10'-0" (1524 x 3048) **landing #3** at the lower left-hand corner of the lobby (Figure 2.7).

8) Use **Draw** > **Ramp** to draw a roughly 14'-0" x 5'-0" (4267 x 1524) ramp #2 connecting **landing #2** and **landing #3** (Figure 2.8). Pay attention to the **direction of travel** for the ramp. Use **Zoom** to zoom in, and use Move Adjust to adjust the ramp #2 to the accurate 14'-0" (4267) dimension (Figure 2.9). Since the ramp #2 is 14'-0" (4267) long, the elevation of the landing #3 should be 14" (355) lower than landing #2 = 26"- 14" = 12" (660-355 = 305).

9) Click on **Set Elevation** to bring up the elevation dialogue box, use the up arrow to set the **landing #3** elevation to 12" (305, Figure 2.10).

10) Use the steps similar to the previous steps 4) to 9) and draw the remaining **landing #4**, **ramps #3 and #4**, and set the elevation of landing #4 to 3" (76, Figure 2.11). Pay attention to the **direction of travel** for the ramps.

11) Since the maximum riser height shall be 7 inches (178) and minimum riser height shall be 4 inches (102). We are going to use 6" (152) high riser, and we need 30"/6" = 5 (762/152 = 5) risers, or 4 treads. We use 11" (279) deep treads, so the total stair length = 4 x 11" = 44" or 3'-8" (4 x 279.5 = 1118). Use **Draw > Stair > Direction (of Stair) > # of Risers > 5** to draw the stair. Pay attention to the **direction of travel** for the stair (Figure 2.12).

12) Use **Draw > Railing** to draw all the railings (Figure 2.13).

Do NOT forget to extend the ramps' non-continuous handrails horizontally at least **12 inches** (305) beyond the **top and bottom** of the ramp run. Do NOT forget to extend the stair's non-continuous handrails horizontally at least **12 inches** (305) beyond the **top and bottom** risers. The railings at the corner of the upper exit corridor have to turn 90 degrees and extend the **12 inches** (305).

Note: Per NCARB program, there is an exception: Handrails are not required on ramps where the vertical rise between landings is 6 inches (152) or less. Therefore, the east handrail (but NOT the west handrail) for the ramp #4 and the east handrail for the ramp #1 are not required. The west handrail for the ramp #4 is still required because open sides of landings, floor surfaces, ramps, and stairways shall be protected by a continuous guardrail. If you do not remember which handrails can be omitted, you can show them and you will pass also. We choose to still show these handrails to match the NCARB sample passing solution. Make sure you turn the end of the west stair handrail 90° at the end to maintain the exit width.

13) You can use **Sketch > Rectangle** to draw some temporary rectangles to assist you in locating the end of the 12" (305) horizontal extension. Use **Zoom** to zoom in and **Move, Adjust** to fine tune the railings.

14) Use **Sketch > Rectangle** to draw a 5'-0" x 6'-0" (1524 x 1829) rectangle at the end of the upper exit corridor to locate the full-height wall.

15) Use **Draw > Full Height Wall** to draw a full height wall at the upper exit corridor (Figure 2.14).

16) Use **Sketch > Rectangle** to draw a 2'-0" x 2'-0" (610 x 610) clearance space at the left side of the wall.

17) Use **Draw > Door** to draw a 36"-wide door at the full height wall at the upper exit corridor (Figure 2.15). Make sure the door swing matches the direction of egress.

Note: NCARB has its own rules, and general exit width requirements (like 44" minimum for exit width) do not apply to doors.

18) Use **Zoom** to zoom out, and use **Sketch > Hide Sketch Elements** to hide sketch elements (Figure 2.16)

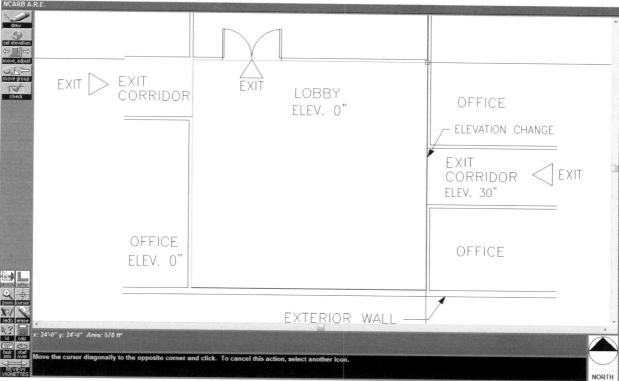

Figure 2.1   Use **Sketch > Rectangle** to draw a rectangle covering the entire lobby.

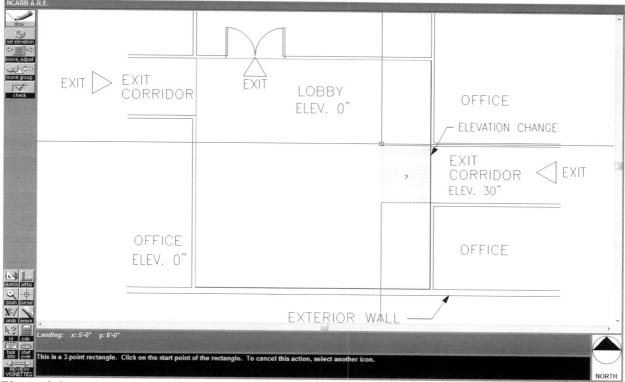

Figure 2.2   Use **Draw > Landing** to draw a landing, which covers the entire width of the exit corridor of the upper level.

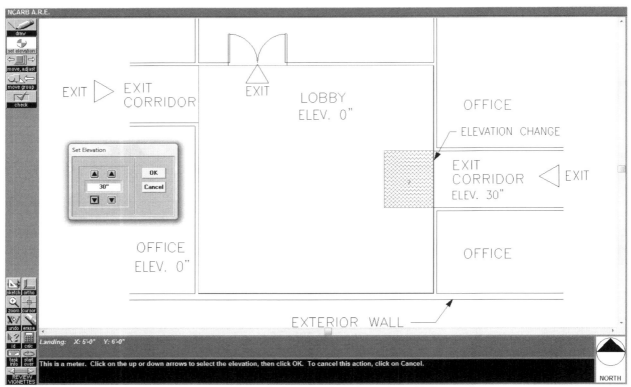

Figure 2.3   Click on **Set Elevation** to bring up the elevation dialogue box.

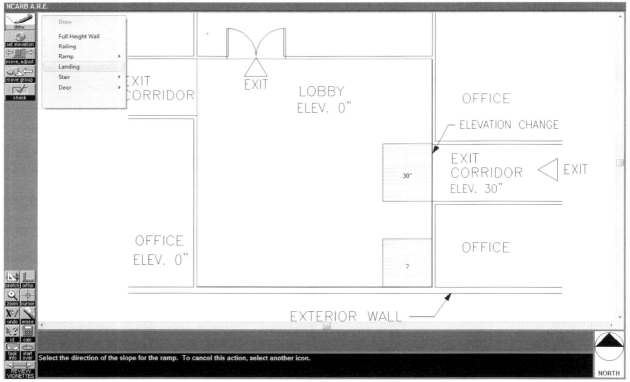

Figure 2.4   Use **Draw > Landing** to draw a 5'-0" x 5'-0" (1524 x 1524) landing at the lower right-hand corner of the lobby.

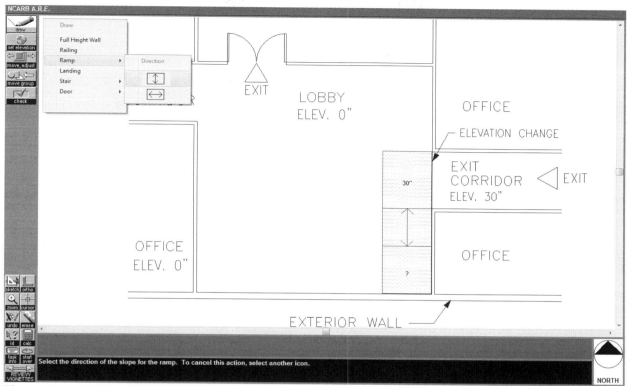

Figure 2.5   Use **Draw > Ramp** to draw a 4'-0" x 5'-0" (1219 x 1524) ramp #1 connecting **landing #1** and landing #2.

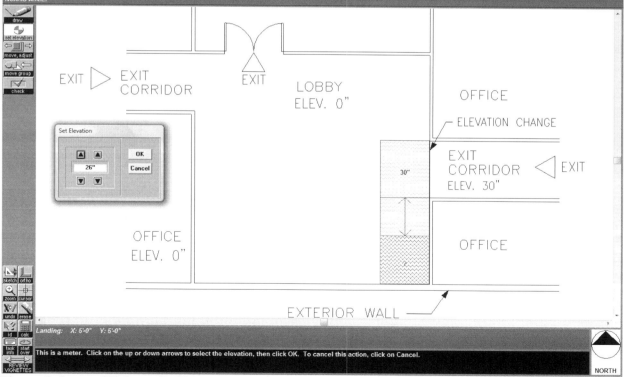

Figure 2.6   Click on **Set Elevation** to bring up the elevation dialogue box (Figure 2.6), use the up arrow to set the **landing #2** elevation to 26" (660).

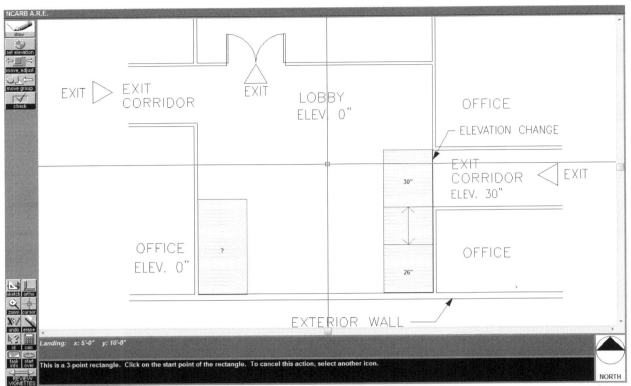

Figure 2.7   Use **Draw > Landing** to draw a 5'-0" x 10'-0" (1524 x 3048) **landing #3** at the lower left-hand corner of the lobby.

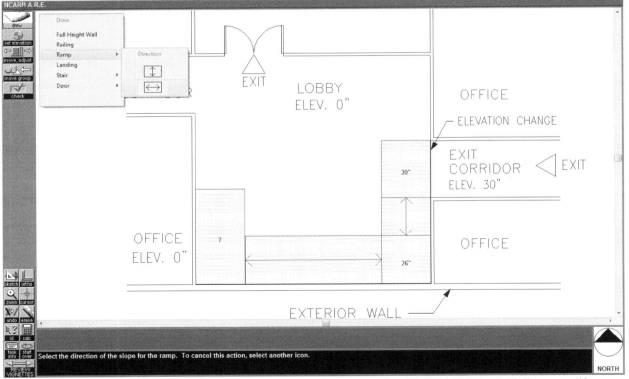

Figure 2.8   Use **Draw > Ramp** to draw a roughly 14'-0" x 5'-0" (4267 x 1524) ramp #2 connecting **landing #2** and **landing #3**.

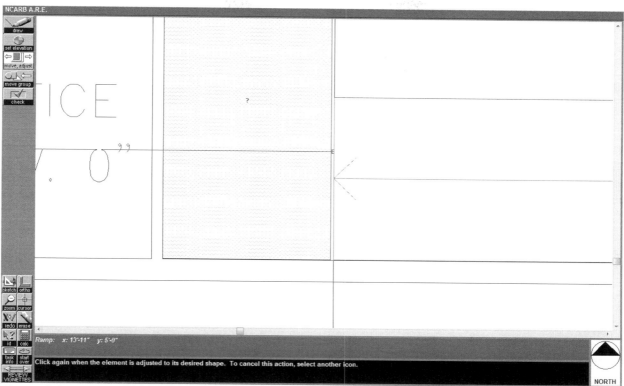

Figure 2.9   Use **Zoom** to zoom in, and use **Move, Adjust to** adjust the ramp #2 to the accurate 14'-0" (4267) dimension.

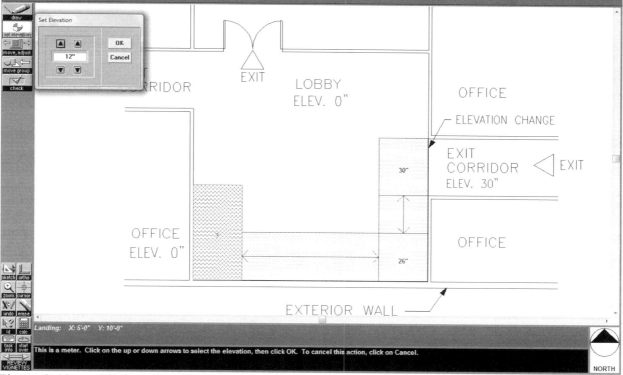

Figure 2.10   Click on **Set Elevation** to bring up the elevation dialogue box.

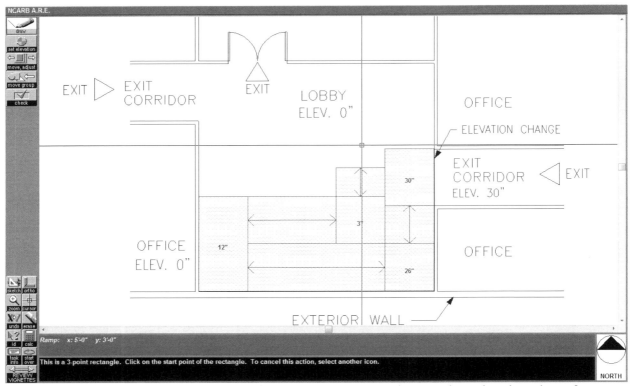

Figure 2.11   Draw the remaining landing, #4, ramps #3 and #4, and set the elevation of landing #4 to 3" (76).

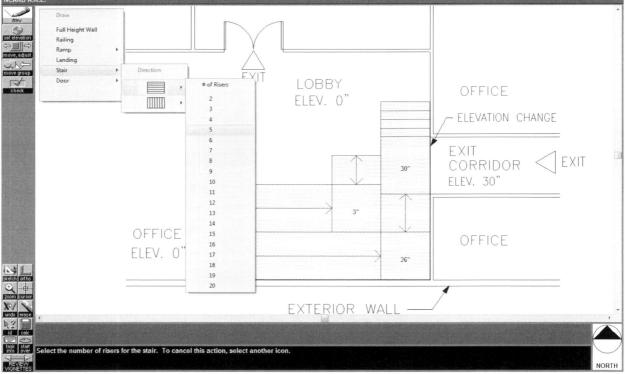

Figure 2.12   Use **Draw > Stair > Direction (of Stair) > # of Risers > 5** to draw the stair.

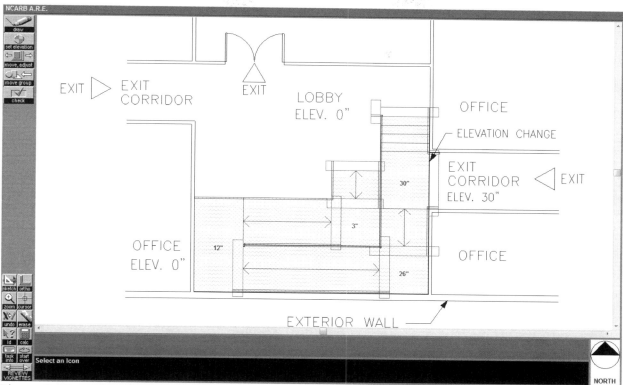

Figure 2.13   Use **Draw >Railing** to draw all the handrails and guardrails.

Figure 2.14   Use **Draw > Full Height Wall** to draw a full height wall at the upper exit corridor.

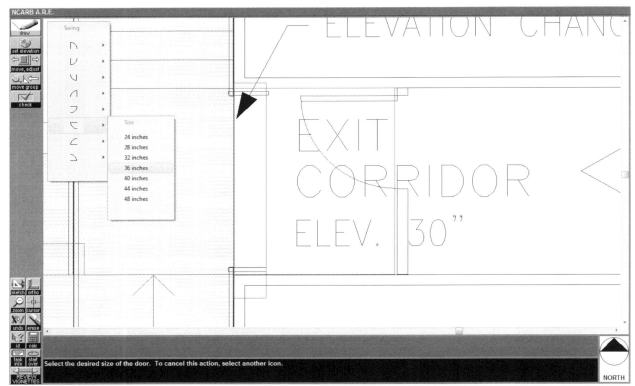

Figure 2.15   Use **Draw > Door** to draw a door at the full height wall at the upper exit corridor.

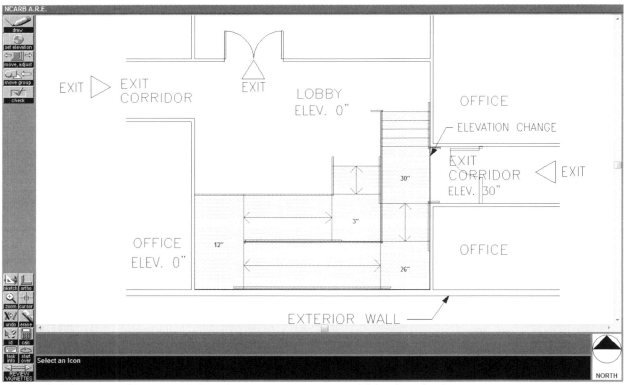

Figure 2.16   Use **Sketch > Hide Sketch Elements** to hide sketch elements.

3. Notes on NCARB traps
   Several **common errors** or **traps** into which NCARB wants you to fall:
1) The **width of the ramp** is less than **5'-0" (1524)**.
2) The minimum dimension for the **landing** is less than **5'-0" (1524)**.
3) You miss the **elevations** for the landings.
4) Forgetting to extend the ramp's non-continuous handrails horizontally at least **12 inches** (305) beyond the **top and bottom** of the ramp run.
5) Forgetting to extend the stairs' non-continuous handrails horizontally at least **12 inches** (305) beyond the **top and bottom** risers.
6) Forgetting to **turn railings at the corner of the upper exit corridor** 90 degrees and extend the **12 inches** (305) horizontally.
7) Forgetting to draw the **full-height wall** and the related door
8) Exit door swings in the wrong direction.
9) Not enough clearance for the pull side of the egress door.

4. A summary of the critical dimensions

1) The **minimum** dimension for the **landing**:	5'-0" (1524)
2) **Ramp width**:	5'-0" (1524)
3) **Stair width**:	5'-0" (1524)
4) **Treads**:	4 @ 11" = 44" (4 @ 279.5 = 1118)
5) **Risers**:	5 @ 6" = 30" (5 @ 152 = 760)
6) Handrail **extension** at **top and bottom** of the ramp runs/Stair:	12" (127)
7) **Clearance** for the **pull side** of the egress door:	24" and 5'-0" x 5'-0" (610 and 1524 x 1524)
8) **Door Size**:	3'-0" (914)

5. An alternate and simple solution to the official NCARB vignette

Figure 2.17 shows an alternate and simple solution to the official NCARB vignette.

Note: Make sure you turn the end of the west stair handrail 90° at the end to maintain the exit width: The width between the west stair handrail and north wall should be no less than the stair width.

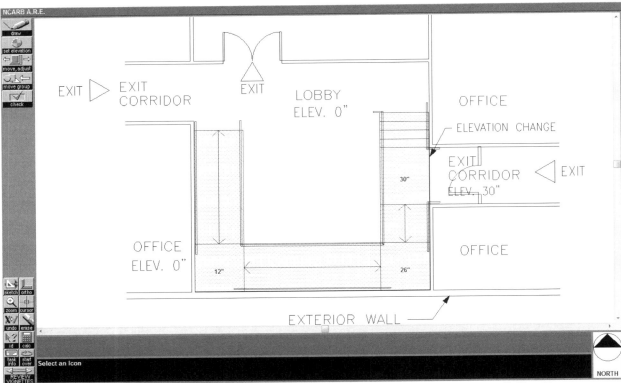

Figure 2.17   An alternate and simple solution to the official NCARB Accessibility/Ramp vignette.

E.   Stair Design Vignette:

1.   Major criteria, overall strategy and tips for stair design vignette:

1) Show proper elevations of all landings.
2) Show proper elevations of the top and bottom of all stair runs.
3) Do NOT forget to extend the ramp's non-continuous handrails horizontally at least **12 inches** (305) beyond the **top and bottom** risers.
4) Make sure there is adequate headroom (80" or 2032).
5) The stair should NOT block the egress door and the lobby door on the ground floor.
6) Minimum stair width 44" (1118), subject to occupant load width increase.
7) Show the 30" x 48" (762 x 1219) area of refuge, and the 24" (610) and 5'-0" x 5'-0" (1524 x 1524) clearance space at the pull side of the egress door. Stair width increased to **48" (1219) clearance between handrails** when area of refuge is required.
8) The width of a landing shall NOT be less than the width of the stair.
9) Thickness of structure under the stair is 12" (1219).
10) Start the stair design from the second floor.
11) By reviewing the ground floor plan and second floor plan, we decide to run the major stair runs along the right wall, and then along the bottom wall to avoid the door at lobby door on the ground floor, and to provide adequate headroom for the exit door on the

ground floor. This layout will also make it possible for the janitor room to use the same stair for access.

2. Step-by-step solution for the official NCARB BDCS practice program stair design vignette:

1) Setting the **stair width** and **minimum landing width:**
- Minimum stair width: 44" (1118)

- The exit width determined by the calculation based on occupant load:

Building Level	Total Occupant Load	Number of Exits	Exit Width
Ground Floor	360	3	(360/3) x 0.3 = 36" (914)
Janitor	9	1	9 x 0.3 = 2.7" (69)
Second Floor	180	2	(180/2) x 0.3 = 27" (686)

Per NCARB program, the exit width shall be based on the individual floor with the largest occupant load, so, the exit width determined by the calculation based on occupant load is 36" (914). This means the **occupant load** of each level will not require a stair width larger than the 44" (1118) minimum.

- Since we do have an **area of refuge** for this vignette, the stair width needs to be at least 56" or 1422 (48" or 1219 CLEAR because of the area of refuge, **plus** an 8" or 203 allowance for handrails on both sides). We set the **stair width** and **minimum landing width** as **5'-0"** (1524) to accommodate the railing, and simplify the risers and treads calculations. Setting the stair width to 5'-0" (1524) give 2" (51) extra on each side of the stair, and give us more room to draw the railings.

Note: NCARB throws in the 44" (1118) minimum stair width requirement at the beginning of the program, and place the 48" (1219) CLEAR stair width requirement because of the area of refuge at the end of the program. This is to make sure you can coordinate various criteria, and make sure you are patient enough to read the entire program.

2) We start the stair design from the second floor:
Use **Layer > Floor Selection > 2** to only turn on the second floor layers (Figure 2.18).

3) Use **Sketch > Rectangle** to draw a 5'-0" x 5'-0" (1524 x 1524) rectangle to show the clearance space for the second floor door.

4) Use **Sketch > Rectangle** to draw a 2'-6" x 4'-0" (762 x 1219) rectangle to show the 30" x 48" (762 x 1219) area of refuge near the second floor door (Figure 2.19).

5) Use **Draw > Landing** to draw an 8'-0" x 6'-0" (x: 8'-0", y: 6'-0") or 2438 x 1829 (x: 2438, y: 1829) landing #1 on the upper-right hand corner of the space to cover both the area of refuge and the 5'-0" x 5'-0" (1524 x 1524) clearance space for the second floor door (Figure 2.20).

6) Use **Draw > Landing** to draw a 5'-0" x 5'-0" (1524 x 1524) landing #2 on the lower-right hand corner of the space (Figure 2.21).

7) Calculate **how many risers we need for the stair runs**:
Per the NCARB vignette requirements, the elevation at the janitor level is 1'-9" or 21" (522), and ground floor elevation is 0'-0". The elevation at the second floor is 12'-3" or 147" (3734).

The riser height has to be 7" (178) maximum, and 4" (102) minimum. Let us use 7" (178) high risers:

147"/7 = **21 risers** needed between the first floor and the second floor.
21"/7 = **3 risers** needed between the first floor and the janitor level.

8) Estimate **how many 11" deep treads we can fit** between landing #1 and landing #2:
Total room vertical length- landing #1 - landing #2 = (18'-8") - (6'-0") – (5'-0") = 7'-8" = 92" = **length available for placing treads (2337)**.

Number of treads needed = length available for placing treads/11" = 92"/11" = 8.36 or 8

Therefore, we are going to place 8 treads or 9 risers between landing #1 and landing #2.

The total length of 8 treads = 8 x11 = 88" or 7'-4" (2235)
The total height of 9 risers = 9x7 = 63" or 5'-3" (1600)

So, landing #2 is 5'-3" (1600) lower than landing #1, its elevation = (12'-3") – (5'-3") = **7'-0"** (2134)

9) Use **Draw > Stairs** to draw a 7'-4" (2235) long stair #1 with 9 risers above landing #2. Pay attention to the direction of travel for the stair (Figure 2.22).

10) Use **Zoom** to zoom in, and **Draw, Adjust** to adjust the landing size to align with the top of stair #1 (Figure 2.23).

11) Click on **Elevation** to bring up the elevation dialogue box, use the **up arrow** to set the elevation of landing #1 to 12'-3" (3734) to match the elevation of second floor.

Set the elevation of top of stair #1 to 12'-3" (3734) to match the elevation of landing #1.

Set the elevation of landing #2 to 7'-0" (2134). See step 8) for calculation of elevation of landing #2.

Set the elevation of bottom of stair #1 to 7'-0" (2134) to match the elevation of landing #2 (Figure 2.24).

Note: Since the structural thickness for the landing and the stair run is 12" or 1'-0", so the clearance under the landing #1 is (12'-3") – (1'-0") = 11'-3" > 8' (the door height on the ground floor) > 80" headroom required.

12) The number of **risers** needed between the janitor level and landing #2 = Total number of **risers** needed between the first floor and the second floor - number of **risers** needed between the first floor and the janitor level - The number of **risers** between the landing #1 and landing #2 = 21-3-9 = 9

See step 8) for previous calculations.

A stair with 9 risers has 8 treads.

The total length of 8 treads = 8 x11 = 88" or 7'-4" (2235)
The total height of 9 risers = 9x7 = 63" or 5'-3" (1600)

13) Use **Draw > Stairs** to draw a 7'-4" (2235) long stair #2 with 9 risers to the left of landing #2. Pay attention to the direction of travel for the stair (Figure 2.25).

*Note: Since all information about the stairs can be shown on the second floor, we chose not to use the **Cut Stair** tool for this solution.*

*In the later part of this book, we provide an alternate solution that does use the **Cut Stair** tool. You need to become familiar with the **Cut Stair** tool because there is a good chance you will be using it in the real exam.*

14) Use **Draw > Landing** to draw a 5'-0" (1524) wide landing #3 on the lower-left hand corner of the space (Figure 2.26).

15) Click on **Elevation** to bring up the elevation dialogue box, use the **up arrow** to set the elevation of landing #3 to 1'-9" (522) to match the elevation of janitor room.

Note: There is a defect in NCARB software. If you set the elevation to 1', and then try to set the 9", you may NOT be able to set the elevation to 1'-9". You may have to set the elevation to 0', and then keep clicking on the up arrow for inch to set the elevation to 1", 2"... and then all the way to 1'-9".

Set the elevation of top of stair #2 to 7'-0" (2133) to match the elevation of landing #2.

Set the elevation of bottom of stair #2 to 1'-9" (522) to match the elevation of landing #3 (Figure 2.27).

16) Use **Draw > Stairs** to draw a 1'-10" (559) long stair #3 with 3 risers (2 treads) to the top of landing #3. Pay attention to the direction of travel for the stair (Figure 2.28).

17) Click on **Elevation** to bring up the elevation dialogue box, use the **up arrow** to set the elevation of top of stair #3 to 1'-9" (522) to match the elevation of landing #3.

Set the elevation of bottom of stair #3 to 0'-0" to match the elevation of ground floor (Figure 2.29).

18) Use **Draw > Railing** to draw all the railings (Figure 2.30).

Do NOT forget to extend the ramp non-continuous handrails horizontally at least **12 inches** (305) beyond the **top and bottom** risers.

You can use **Sketch > Rectangle** to draw some temporary rectangles to assist you to locate the end of the 12" (305) railing extensions.

Use **Zoom** to zoom in, and **Move, Adjust** to fine turn the railings (Figure 2.31).

19) After you finish all railings, Use **Zoom** to zoom out, and use **Sketch > Hide sketch elements** to hide the sketch elements. This is your final solution (Figure 2.32).

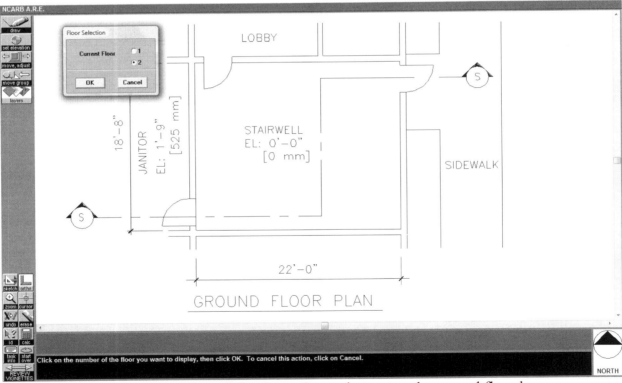

Figure 2.18   Use **Layer > Floor Selection > 2** to only turn on the second floor layers.

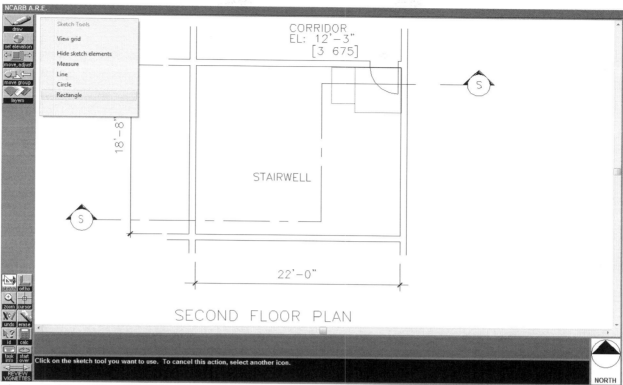

Figure 2.19   Use **Sketch > Rectangle** to draw a 5'-0" x 5'-0" (1524 x 1524) rectangle to show the clearance space for the second floor door.

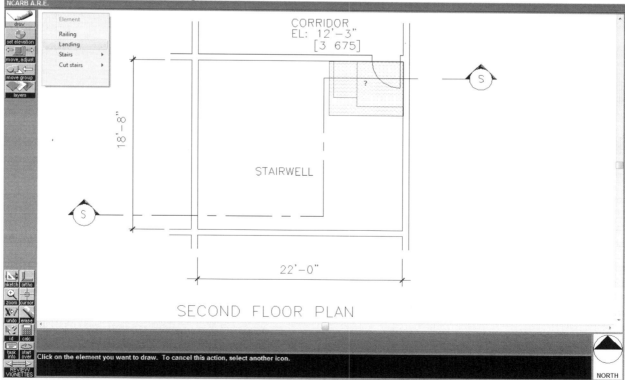

Figure 2.20   Use **Draw > Landing** to draw an 8'-0" x 6'-0" (2439 x 1829) landing #1.

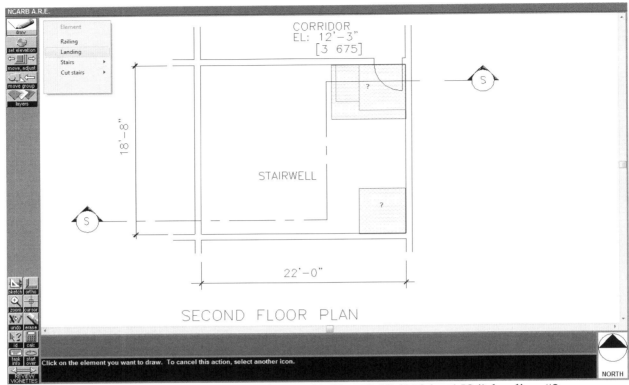

Figure 2.21   Use **Draw > Landing** to draw a 5'-0" x 5'-0" (1524 x 1524) landing #2.

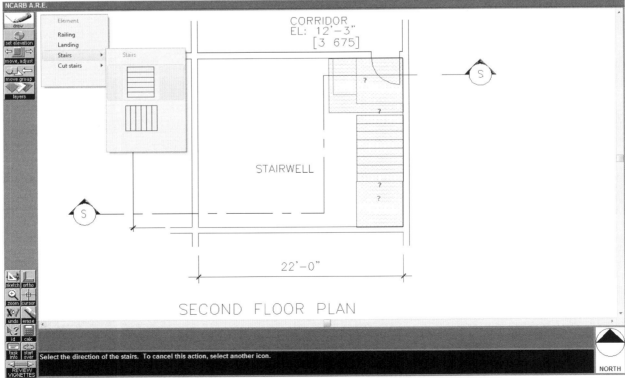

Figure 2.22   **Draw > Stairs** to draw a 7'-4" (2235) long stair with 9 risers above landing #2.

Figure 2.23   Use **Zoom** to zoom in, and **Draw, Adjust** to adjust the landing size to align with the top of stair #1.

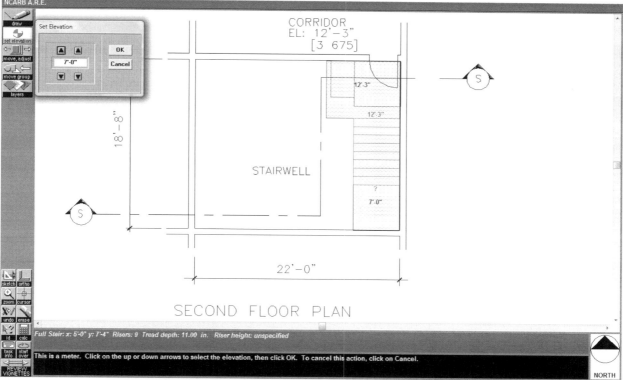

Figure 2.24   Clock on **Elevation** to bring up the elevation dialogue box, use the **up arrow** to set the elevation of landing #1 to 12'-3" (3734).

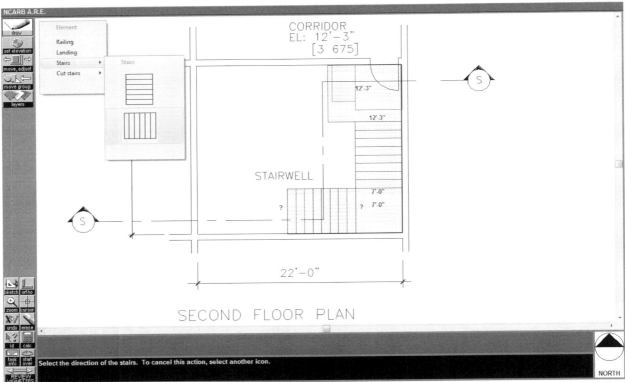

Figure 2.25   Use **Draw > Stairs** to draw a 7'-4" (2235) long stair #2 with 9 risers to the left of landing #2.

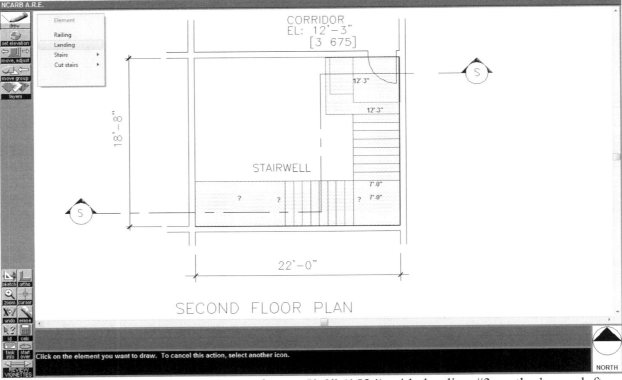

Figure 2.26   Use **Draw > Landing** to draw a 5'-0" (1524) wide landing #3 on the lower-left hand corner of the space.

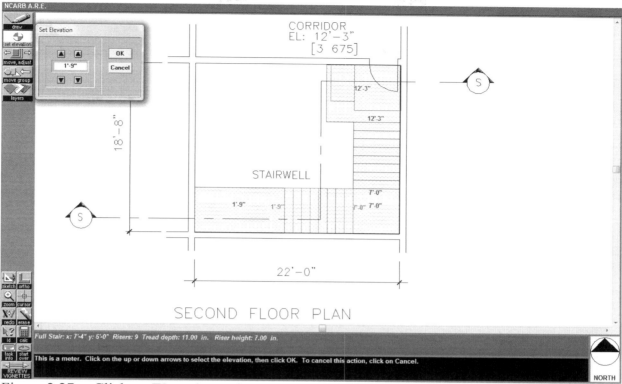

Figure 2.27   Click on **Elevation** to bring up the elevation dialogue box, use the **up arrow** to set the elevation of landing #3 to 1'-9" to match the elevation of janitor room.

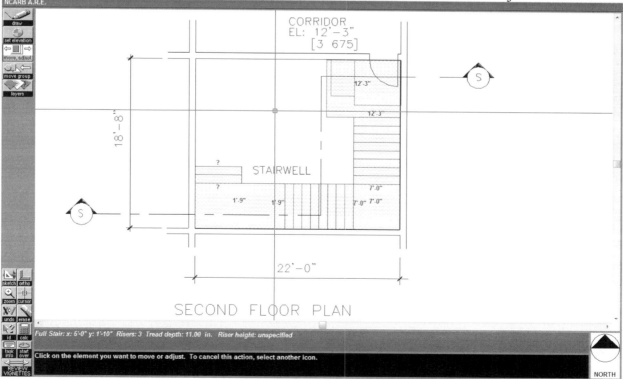

Figure 2.28   Use **Draw** > **Stairs** to draw a 1'-10" (559) long stair #3 with 3 risers (2 treads) to the top side landing #3.

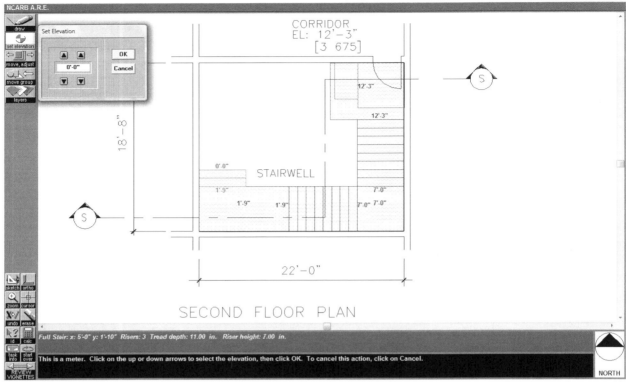

Figure 2.29   Click on **Elevation** to bring up the elevation dialogue box, use the **up arrow** to set the elevation of top of stair #3 to 1'-9" to match the elevation of landing #3.

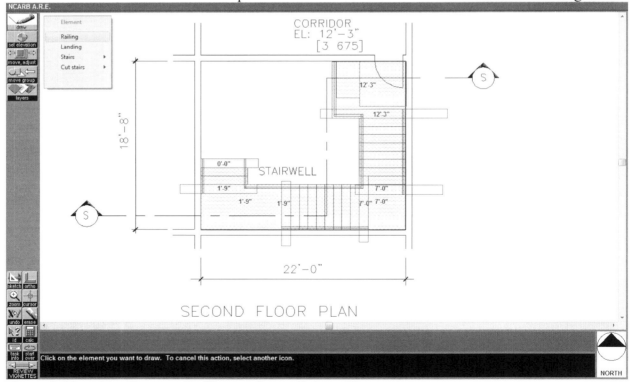

Figure 2.30   Use **Draw > Railing** to draw all the railings.

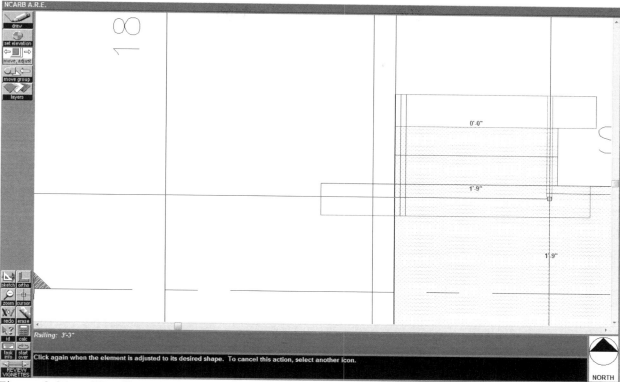

Figure 2.31   Use **Zoom** to zoom in, and use **Move, Adjust** to fine turn the railings.

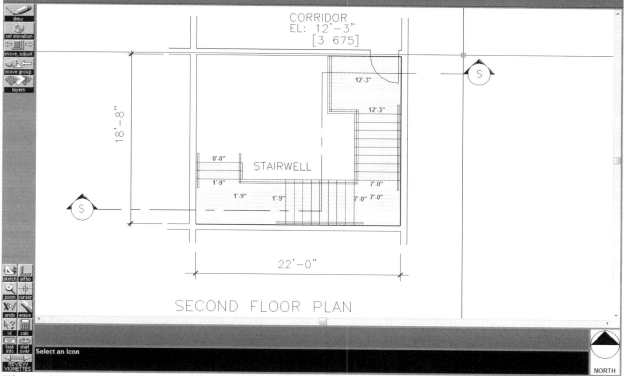

Figure 2.32   Final solution: use **Sketch > Hide sketch elements** to hide the sketch elements.

3. Notes on NCARB traps

Several **common errors** or **traps** into which NCARB wants you to fall:

1) Forgetting to show **proper** elevations of all landings.
2) The elevation of top or bottom of the stair run **does not match** the adjacent elevation of the landing or floor.
3) Forgetting to extend the ramp non-continuous handrails horizontally at least **12 inches** (305) beyond the **top and bottom** risers.
4) There is not enough **headroom (80"or 2032)** for the doors under the landing or stairs.
5) Stairs block the egress door or the lobby door on the ground floor.
6) Forgetting to show the 30" x48" (762 x 1219) area of refuge, or the 24" (610) and 5'-0" x 5'-0" (1524 x 1524) clearance space at the pull side of the egress door. Forget stair width increase to **48" (1219) clear between handrails** when area of refuge is required. This can happen if you do NOT read ALL the NCARB requirements.
7) The width of a landing is less than the width of the stair.
8) Design two stairs instead of ONE stair as required by the NCARB program.
9) Riser height is not the same throughout the stair runs.
10) The number of risers and riser heights do NOT match or add up to the elevations for the janitor level or the second floor level.

4. A summary of the critical dimensions

1) The **minimum** dimension for the **landing:**	5'-0" (1524)
2) **Stair width (4'-8" minimum):**	5'-0" (1524)
3) **Treads**:	11" (279.5)
4) **Risers**:	7" (178)
5) Handrail **extension** at **top and bottom** of the stair runs:	12" (305)
6) **Clearance** for the **pull side** of the egress door:	24" and 5'-0" x 5'-0" (610 and 1524 x 1524)
7) **Refuge area**:	30"x48" (762 x 1219)
8) **Headroom**:	80" (2032)
9) Ground level exit **door height**	8'-0" (2438)
10) **Clearance** under the second floor landing	11'-3" (3429)

(Adequate to clear the headroom and the ground level exit door below**)**

5. Practice using the cut stair tool. You will probably need to use it in the real exam

Practice using the cut stair tool. You will probably need to use it in the real exam. It is better to know how to use the cut stair tool, and not have to use it in the real exam, than the other way around.

You need to show the first floor information on the first floor, and the second floor information on the second floor. The stair between the first and second floor will be "cut" into half.

Several **key points**:
1) When using the cut stair tool, draw the flight of the cut stairs from landing to landing or from landing to the ground floor.
2) The flight of the cut stairs has to show up on both the first and the second floor. You need to draw the same flight twice: once for the first floor and the other for the second floor.
3) Make sure ALL the data for the same flight of the cut stairs is the same for both floor: top and bottom elevations, depths of treads, and number of risers.
4) Extend the railing 1 to 2 clicks over the break line.

The following are the steps to redraw the previous solution using the cut stair tool.
1) Erase the stair #2, #3 and landing #3 as well as the related railing from the second floor.

2) Use **Layers > Current floor > 1** to set the current floor to the first floor, and redraw the stair #3 and landing #3 on the first floor (Figure 2.33):
   - Landing #3: X = 9'-8" (2946), Y = 5'-0" (1524), Elevation = 1'-9" (533)
   - Stair #3: Width = 5'-0" (1524), Length = 2 Treads @11" = 1'-10" (559), 3 Risers at 7" = 21" (533)

See steps 12) to 17) in the original solution for details on the above data.

3) Draw the cut stair for the first floor:
   - The data for stair #2: Width = 5'-0" (1524), Length = 8 Treads @11" =7'-4" (2235), 9 Risers at 7" = 63" (1600)
   Use **Draw > Cut Stair > (Select the proper cut stair symbol)** to draw the cut stair for the first floor. Pay attention to the orientation of the cut stair symbol. Set the riser number to 9, and set the top and bottom of the cut stair elevation to match the adjacent landing elevations (Figure 2.34).

4) Draw the railing for stair #3, cut stair #2, and landing #3. Make sure to extend the railing 1 to 2 clicks over the break line for the cut stair (Figure 2.35).

5) Use **Layers > Current floor > 2** to set the current floor to the second floor.

6) Draw the cut stair for the second floor:
   Use **Draw > Cut Stair > (Select the proper cut stair symbol)** to draw the cut stair for the second floor. Pay attention to the orientation of the cut stair symbol. Set the riser number to 9, and set the top and bottom of the cut stair elevation to match the adjacent landing elevations.

7) Draw the railing for cut stair #2. Make sure to extend the railing 1 to 2 clicks over the break line (Figure 2.36).

8) Your alternate solution for using the cut stair tool is now finished (Figure 2.35 & Figure 2.36).

6. A much tougher and very helpful stair vignette

The **old** ARE 3.1 version has a much tougher stair vignette. Based on the readers' feedbacks for the **new** ARE 4.0 exam BDCS division, you have a good chance of getting a version of the stair vignette that is very similar and is as difficult as the old ARE 3.1 version.

See the following link for the FREE download of the NCARB ARE 3.1 exam guide, practice vignettes and graphic software:
http://www.ironwarrior.org/ARE/_NCARB_Software/3_1/

I highly recommend you download the information, try out the old stair vignette and read the sample solutions in the file entitled "Graphics_31.pdf." You will be very happy you did. It WILL definitely help you in the real exam and current version of the ARE BDCS division graphic vignettes.

This old ARE 3.1 vignette is a great way for you to exercise the **Cut Stair** command of the NCARB graphic software. If you have more time, check out the old ARE 3.1 version of the roof plan vignette and other vignettes also.

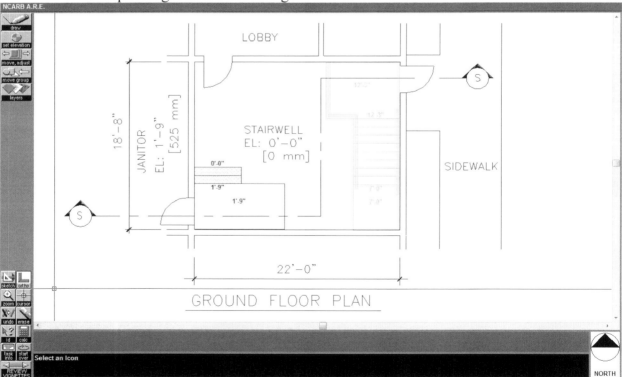

Figure 2.33   Use **Layers > Current floor > 1** to set the current floor to the first floor, and redraw the stair #3 and landing #3 on the first floor.

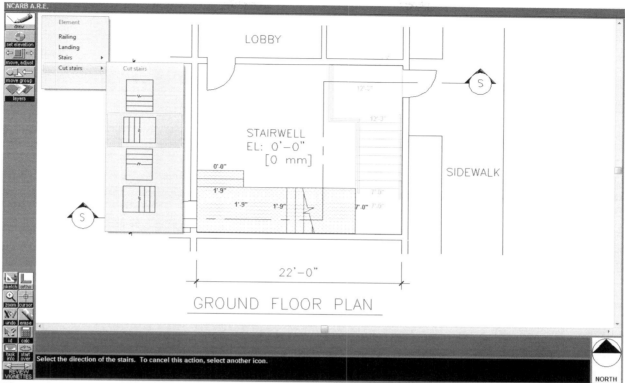

Figure 2.34   Use **Draw > Cut Stair > (Select the proper cut stair symbol)** to draw the cut stair for the first floor.

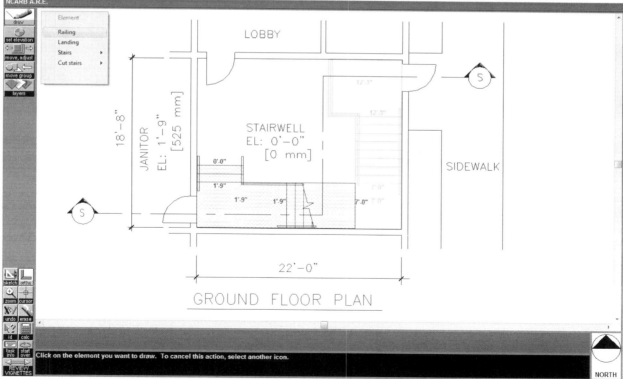

Figure 2.35   Draw the related railing for stair #3, cut stair #2, and landing #3. Make sure to extend the railing 1 to 2 clicks over the break line.

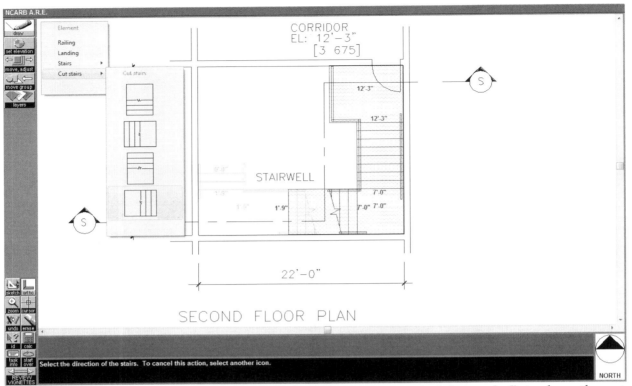

Figure 2.36   Use **Draw > Cut Stair > (Select the proper cut stair symbol)** to draw the cut stair for the second floor.

F. Roof Plan Vignette:

1. Major criteria, overall strategy and tips for roof plan vignette:
1) The building has one high roof and one low roof.
2) Outside edges of the roof planes must coincide with the dashed lines, indicating the outermost edges of the roofs. Gutters and downspouts can be placed beyond the dashed lines.
3) The slope for the roof over the exhibition room shall be between 6:12 and 12:12.
4) The slope for the roof over the remaining spaces shall be between 2:12 and 5:12.
5) Check the elevations of the low points of upper roof and high points of the lower roof, and make sure you leave adequate space to accommodate the 1'-6" (457) thick roof and structural assembly and the **2'-0" (610) high clerestory window** in the west wall.
6) Provide skylights for rooms, which have no windows and no clerestory window.
7) Halls, storage rooms, or closets do not need skylights.
8) Check the elevations of the low points of the lower roof, and make sure you leave adequate space to accommodate the 1'-6" (457) thick roof and structural assembly and the **8'-0" (2438) high first floor ceiling**.
9) Do NOT miss the gutters or downspouts.
10) Do NOT miss the cricket for the chimney.
11) Do NOT miss the plumbing vent stacks, and exhaust fan vents for the restrooms and the kitchen.
12) Do NOT miss the skylights at the Men's Restroom and the Women's Restroom.

13) Place the HVAC condensing unit on a roof with a slope of 5:12 or less, but NOT in front of the clerestory window.

14) Make sure the HVAC condensing unit has the required 3'-0" (914) minimum clearance from all roof edges.

2. Step-by-step solution for the official NCARB BDCS practice program roof plan vignette:

1) **Calculating the roof heights:**
The elevation for the lowest points of lower roof = the roof and structural assembly thickness + the first floor ceiling height = (1'-6") + (8'-0") = 9'-6" (2896)

The **minimum** difference between the upper roof and lower roof = the roof and structural assembly thickness + the clerestory window height = (1'-6") + (2'-0") = 3'-6" (1067)

2) Use **Draw > Roof Plan** to draw part of the lower roof (Figure 2.37).

3) Click on **Set roof**, and then click on the **arrow symbol** to set the direction of the slope. The arrow must point in the direction of the **downward** slope. Click on the **question mark** to bring up a dialogue box, use the **up or down arrows** to set the slope ratio to 2:12. Click on **Set roof**, and then click on another **question mark** to bring up a dialogue box, use the up or down arrows to set the elevation for the lowest points of lower roof to 9'-6" (2896, Figure 2.38).

*Note: After you bring up a dialogue box, the **up or down arrow** is **tricky and hard to use**. For example, if you want to set an elevation to 12'-6", if you set the 12', and then try to set the 6", you may NOT be able to do it, you'll get 12'-5" or 12'-7", but NOT 12'-6". You have to set it to 11'-9" first, and then keep clicking on the **up arrow next to the inches** to increase it as 11'-9", 11'-11", 12'-0", 12'-1" and so on until you reach 12'-6". You need to practice and play with the NCARB software and beware of tricky things like this. Otherwise, you will be wasting your valuable time to explore this in the real ARE test.*

4) Draw the remaining lower roof using commands similar to step 2) and 3) (Figure 2.39).

Because we are using 2:12 or 1:6 slope, and the horizontal distance between the low point and the high point of the lower roof is 18', so their difference between their elevations is (18'-0")/6=3'-0" (914)

The elevation of the high point of the lower roof = (9'-6") + (3'-0") = 12'-6" (3810)

5) Use **Zoom** to zoom in every corner of the roof plans, and use **Move, Adjust to** adjust the edge of the roof planes to align with the dashed lines (Figure 2.40).

6) Draw the upper roof using commands similar to step 2) and 3) (Figure 2.41).

We have set the ridgeline in the middle of the upper roof.

The elevation for the lowest points of upper roof = the elevation of the high point of the lower roof + (3'-6") = (12'-6") + (3'-6") = 16'-0" (4879)

The slope of the upper roof is 6:12 or 1:2.

The elevation for the ridgeline of the upper roof = 16' + (22'/2) = 27'-0" (8230)

7) Use the **Check** command to check and make sure that the roof planes align correctly and are NOT overlap. If a roof plane is not drawn correctly, it will become highlighted (Figure 2.42).

8) Use **Zoom** to zoom into every corner of the roof plans, and use **Move, Adjust to** adjust the edge of the roof planes to align with the dashed lines or the lower roof planes if necessary.

9) Use **Draw > Clerestory** to draw the clerestory window in the west wall below the upper roof.

   Note: The roof edges below and above the clerestory must be horizontal and parallel, not sloped.

10) Use **Draw > HVAC condenser** to draw the HVAC condensing unit above the lounge area on the lower roof (Figure 2.43).

   Make sure the HVAC condensing unit has the required **3'-0" (914) minimum clearance** from all roof edges.

11) Because the Men's Restroom and the Women's Restroom have no window or clerestory window, each one needs a skylight.

   Use **Draw > Skylight** to draw a skylight for the Men's Restroom and a skylight for the Women's Restroom.

12) Use **Draw > Exhaust Fan Vent** to draw an exhaust fan for the Men's Restroom and an exhaust fan for the Women's Restroom.

13) Use **Draw > Plumbing Vent Stack** to draw a plumbing vent stack inside the common walls between the Men's Restroom and the Women's Restroom, and the wall behind the kitchen sink (Figure 2.44).

14) Use **Draw > Cricket** to draw a cricket on the slope roof on the high side of the chimney (Figure 2.45).

15) Use **Draw > Gutter** to draw a gutter along the **low** edge of **every** slope roof plane (Figure 2.46).

Note: There is a defect with the gutter command of the NCARB software. Once you place it, it is almost impossible to modify it. There are two ways to solve this problem: One way is ***Zoom out*** *first, and then click on the gutter a few times to select and* ***erase*** *it, and then* ***redraw*** *a new one. The other way is to select the gutter while clicking on the side that you drew it with (usually* ***the side closest to the building***). *You can then move the gutter like any other element. If you want to adjust the gutter length, you need to click on the* ***intersection*** *at the end of the gutter and on* ***the side closest to the building***, *and then adjust it.*

Some people think it is not necessary to show portion of the gutter along the east wall, outside the chimney. I have a different opinion: if you do omit portion of the gutter outside the chimney, then you have to add 2 more downspouts because each segment of the gutter should have at least 2 downspouts, one at each end. This is a common practice: if one of the downspouts is clogged, the water can still drain from the remaining downspout. Therefore, I think it is better to show the portion of the gutter outside the chimney like the suggested solution in NCARB Guide.

16) Use **Draw > Downspout** to draw a downspout on both end of **each** gutter (Figure 2.47).

17) Use **Zoom** to zoom in every corner of the roof plans, and use **Move, Adjust to** adjust the gutters and downspouts (Figure 2.48).

18) Use **Draw > Flashing** to add flashings to **ALL** intersections at walls, roofs and chimney (Figure 2.49).

19) Use **Zoom** to zoom out. Use **Sketch > Hide sketch elements** to hide sketch elements and view the final solution (Figure 2.50).

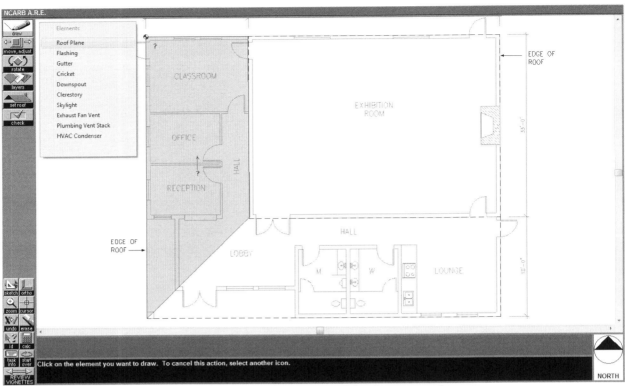

Figure 2.37   Use **Draw > Roof Plan** to draw part of the lower roof.

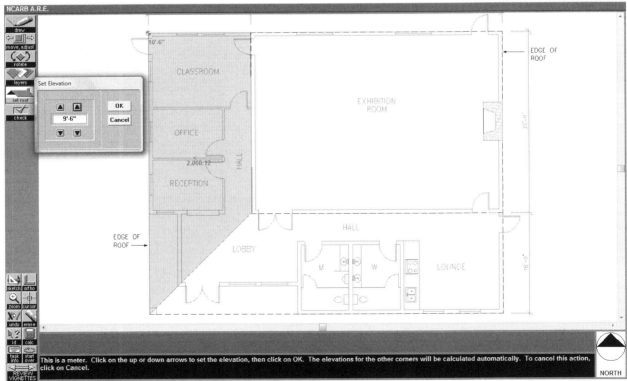

Figure 2.38   Click on the **arrow symbol** to set the direction of the slope, click on a **question mark** to bring up a dialogue box of the elevation.

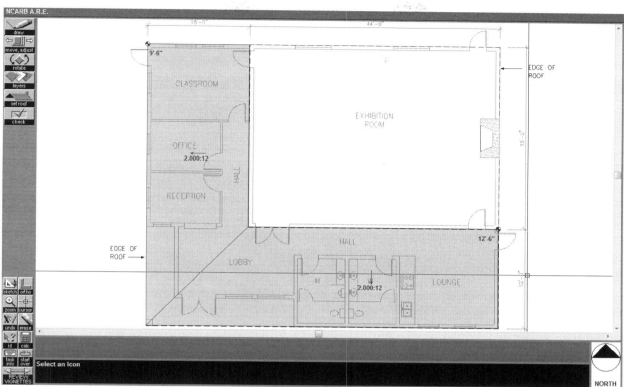

Figure 2.39   Draw the remaining lower roof using similar to step 2) and 3).

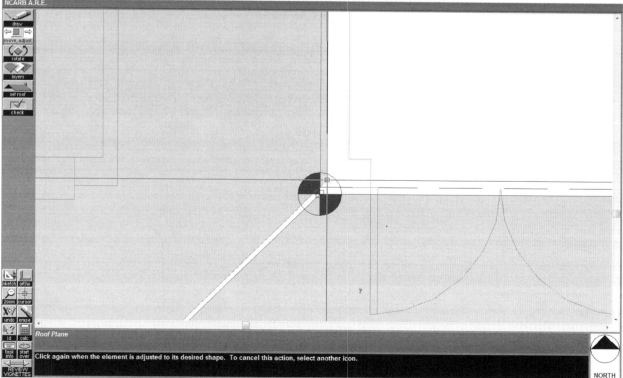

Figure 2.40   Use **Zoom** to zoom in, and use **Move, Adjust to** adjust the edge of the roof to align with the dashed lines.

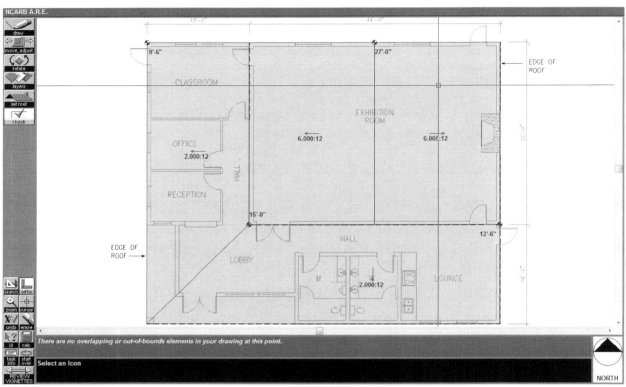

Figure 2.41   Draw the upper roof using commands similar to step 2) and 3).

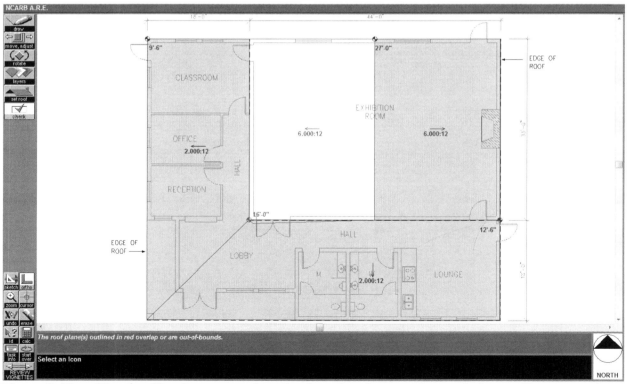

Figure 2.42   Use the **Check** command to check and make sure the roof planes align correctly.

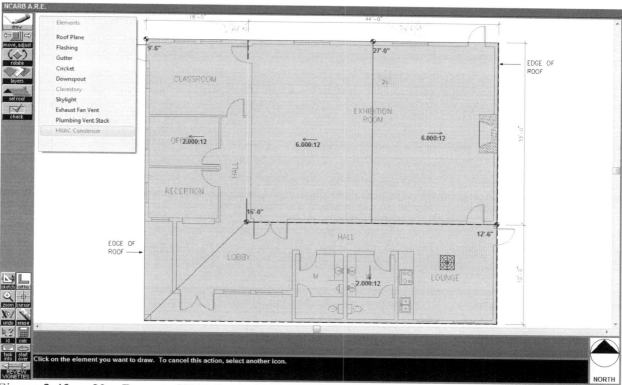

Figure 2.43   Use **Draw > HVAC condenser** to draw the HVAC condensing unit above the lounge area on the lower roof.

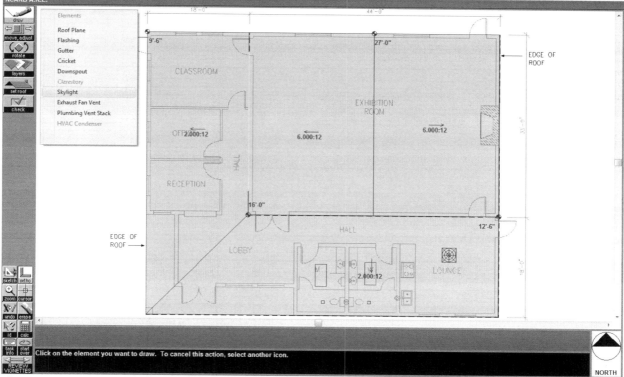

Figure 2.44   Use **Draw > Skylight** to draw a skylight, use **Draw > Exhaust Fan Vent** to draw an exhaust fan, and use **Draw > Plumbing Vent Stack** to draw a plumbing vent.

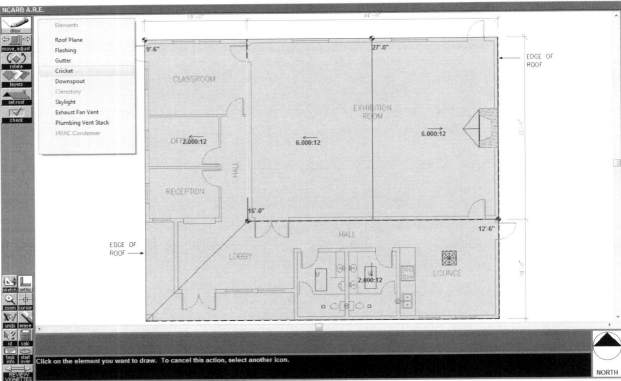

Figure 2.45   Use **Draw > Cricket** to draw a cricket on the slope roof on the high side of the chimney.

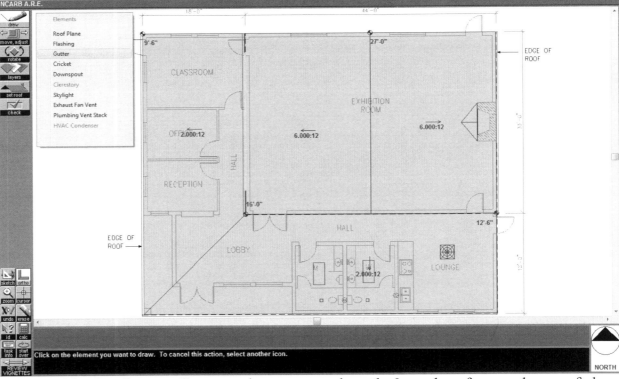

Figure 2.46   Use **Draw > Gutter** to draw a gutter along the **low** edge of **every** slope roof plan.

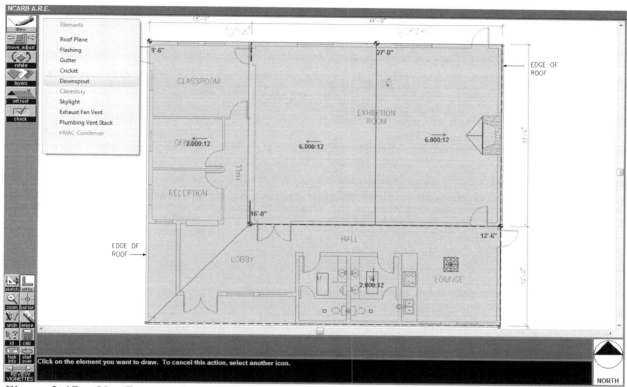

Figure 2.47   Use **Draw > Downspout** to draw a downspout on both end of **each** gutter.

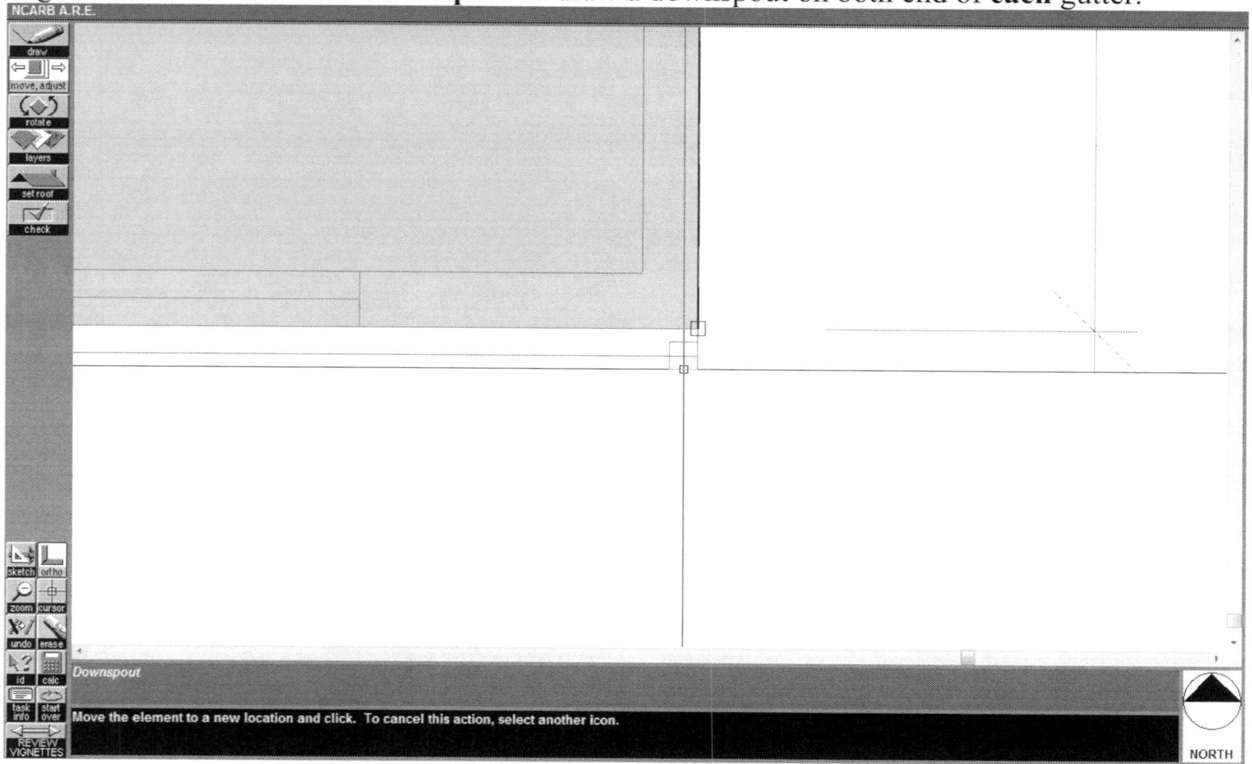

Figure 2.48   Use **Zoom** to zoom in every corner of the roof plans, and use **Move, Adjust to** adjust the gutters and downspouts.

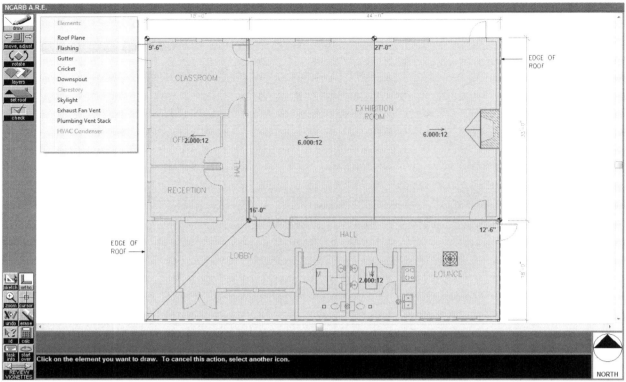

Figure 2.49   Use **Draw > Flashing to** add flashings to **ALL** intersections at walls, roofs and chimney.

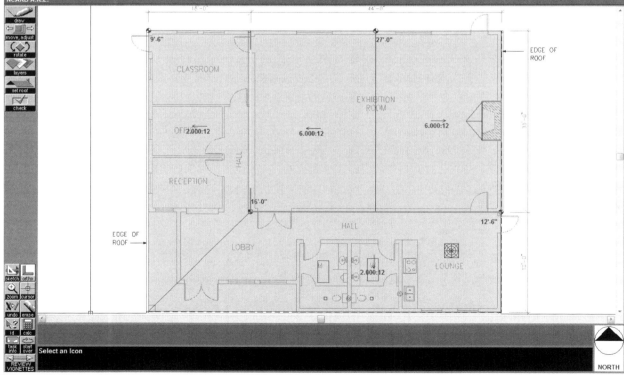

Figure 2.50   Use **Zoom** to zoom out. Use **Sketch > Hide sketch elements** to hide sketch elements and view the final solution.

3. Notes on NCARB traps

Several **common errors** or **traps** into which NCARB wants you to fall:

1) Draw part of the roof planes outside of the **dashed lines**:
   - Outside edges of the roof planes must coincide with the dashed lines indicating the outermost edges of the roofs. Gutters and downspouts can be placed beyond the dashed lines.

2) Show the **wrong slop ratio**:
   - The slope for the roof over the exhibition room shall be between 6:12 and 12:12.
   - The slope for the roof over the remaining spaces shall be between 2:12 and 5:12.

3) Not enough **clearance space** for the roof and structural assembly and **clerestory window:**
   - Check the elevations of the low points of upper roof and high points of the lower roof, and make sure you leave adequate space to accommodate the 1'-6" (457) thick roof and structural assembly and the **2'-0" (610) high clerestory window** in the west wall.

4) Miss **skylights**:
   - Provide skylights for rooms have no windows and no clerestory window.

   - Halls, storage rooms, or closets do not need skylights.

5) Not enough **clearance space** for first floor ceiling:
   - Check the elevations of the low points of the lower roof, and make sure you leave adequate space to accommodate the 1'-6" (457) thick roof and structural assembly and the **8'-0" (2438) high first floor ceiling**.

6) Miss the gutters or downspouts, or the cricket for the chimney, or the plumbing vent stacks, or the exhaust fan vents for the restrooms and the kitchen.

7) Not enough **clearance space** clearance for the HVAC condensing unit:
   - Place the HVAC condensing unit on a roof with a slope of 5:12 or less, but NOT in front of the clerestory window.
   - Make sure the HVAC condensing unit has the required 3'-0" (914) minimum clearance from all roof edges.

4. A summary of the critical dimensions

1) First floor **ceiling** height: 8'-0" (2438)
2) Roof and structural assembly thickness: 1'-6" (457)
3) Low roof:
   - Slope ratio: 2:12
   - Elevation of the **lowest** point: 9'-6" (2896)
   - Elevation of the **highest** point at clerestory window: 12'-6" (3810)
   - Clearance under **lower** roof: 8'-0" (2438)

4) Upper roof:
   - Slope ratio: 6:12
   - Elevation of the **highest** point or the **ridge line**:   27'-0" (8230)
   - Elevation of the **lowest** point at clerestory window: 16'-0" (4877)
   - Clearance under **upper** roof for clerestory window
     and roof structure: (2'-0") + (1'-6") = 3'-6" (1067)

Chapter Three

ARE Mock Exam for
Building Design and Construction Systems (BDCS) Division

A. Mock Exam: BDCS Multiple-Choice (MC) Section

1. Which of the following should be specified to affix shingles onto roof structure? **Check the two that apply.**
   a. Stainless steel nails
   b. Iron nails
   c. Staplers using pneumatic staple guns
   d. Terne-coated stainless steel

2. Which of the following has been banned in new construction? **Check the two that apply.**
   a. Copper nails
   b. Lead-based paint
   c. Zinc-based paint
   d. Asbestos

3. Which of the following should not have direct contact with copper nails?
   a. 4-ply roofing
   b. Fiber-glass roofing
   c. Red-cedar shingles
   d. TPO roofing

Figure 3.1   Lock image

4. What kind of lock is shown on the previous image?
   a. Mortise lock
   b. Unit lock
   c. Integral lock
   d. Cylinder lock

5. Which of the following statements is incorrect? **Check the two that apply.**
   a. Construction joints also serve as isolation or control joints.
   b. Control joints also serve as isolation or construction joints.
   c. Construction joints normally run from the top of slab to the bottom of the slab.
   d. Control joints normally run from the top of slab to the bottom of the slab.

6. The term "fine sand float finish" refers to
   a. plastering
   b. paving
   c. painting
   d. concrete

7. The term "VOC" refers to
   a. Valid Organic Compound
   b. Volatile Organic Compound
   c. Volatile Original Compound
   d. Volatile Original Composite

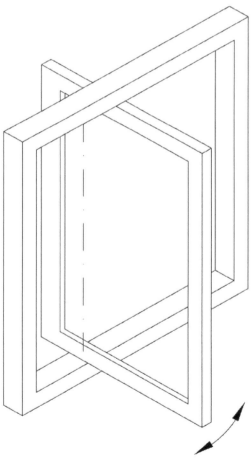

Figure 3.2   Window type

8. What kind of window is shown on the previous image?
    a. Sliding
    b. Pivoting
    c. Casement
    d. Fixed

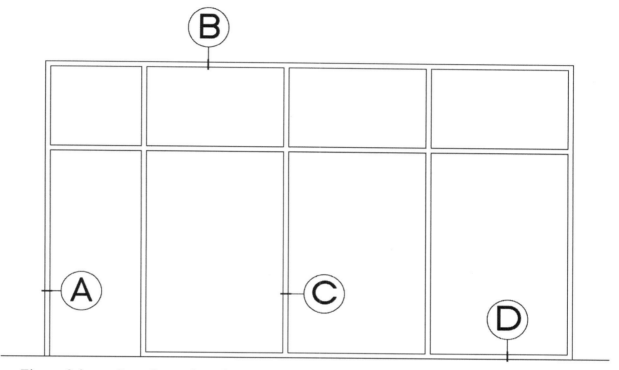

Figure 3.3   Storefront elevation

9. What letter in the previous figure indicates a mullion?
    a. A
    b. B
    c. C
    d. D

10. Which of the following is the most cost-effective finish for exterior walls?
    a. Full-brick veneer
    b. Thin-brick veneer
    c. Stone veneer
    d. Plastering

11. According to *International Building Code* (IBC), which of the following statement is true at stairways where handrails are not continuous between flights?
    a. The handrails shall extend at least 12" (305) beyond the top riser and continue to slope for the depth of one tread beyond the bottom riser.
    b. The handrails shall extend at least 12" (305) beyond the top riser and continue to slope for the depth of one tread plus 12" (305) beyond the bottom riser.
    c. The handrails shall extend at least 12" (305) plus the depth of one tread beyond the top riser and extend at least 12" (305) beyond the bottom riser.
    d. The handrails shall extend at least 12" (305) beyond the top riser and extend at least 12" (305) beyond the bottom riser.

12. According to *International Building Code* (IBC), stairways shall typically have a minimum headroom clearance of _____ measured vertically from a line connecting the edge of the nosing.

13. According to *International Building Code* (IBC), stair riser heights shall be _____ maximum and _____ minimum. Stair tread depths shall be _____ minimum typically.

14. The term "Elastomeric - Modified Acrylic" is most likely to be found in the specifications for
    a. plastering
    b. paving
    c. painting
    d. concrete

15. An architect is doing quality control of a set of construction drawings. The Health Department requires all food prep areas have 8' (2438) high FRP. Where are the best places to find the information for the 8' (2438) high FRP? **Check the two that apply.**
    a. Floor plans
    b. Room finish schedules
    c. Interior elevations
    d. Reflected ceiling plans

16. An architect is drawing the roof plan. There is a chimney penetrating the roof at the central area of the roof. She is trying to drain the rainwater away from the Chimney. Which of the following is true?
    a. She should add gabled flashings on the high side of the roof, next to the chimney, to drain rainwater.
    b. She should add gabled flashings on the low side of the roof, next to the chimney, to drain rainwater.
    c. She should add a cricket on the high side of the roof, next to the chimney, to drain rainwater.
    d. She should add a cricket on the low side of the roof, next to the chimney, to drain rainwater.

17. In a building that uses Insulated Concrete Forms (ICF) system, the forms are left in place permanently for the following reasons: (**Check the four that apply**)
    a. Thermal insulation
    b. Acoustic insulation
    c. Space to run plumbing pipes and electrical conduits.
    d. Aesthetic effect
    e. Construction cost
    f. Backing for gypsum boards, stucco, and brick

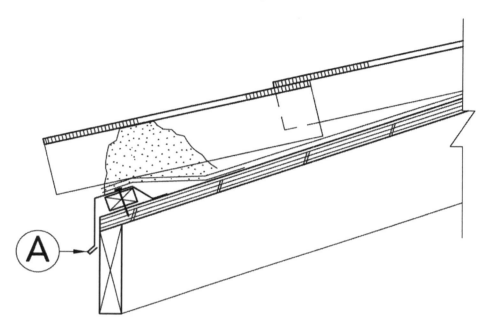

Figure 3.4   Roof detail

18. What is noted as letter A on the previous image?
    a. Flashing
    b. Drip edge
    c. Coping
    d. Siding

19. If a set of building plans has to be reviewed by the Health Department, which of the following is likely to be accepted by the Health Department as a floor finish for the Janitor's room? **Check the three that apply.**
    a. Carpet with wood base
    b. Concrete with slim foot base
    c. VCT flooring with cove base
    d. Sheet vinyl flooring with cove base
    e. Ceramic tile floor with cove base
    f. Smooth wood floor with cove base

20. Which of the following are likely to be most cost-effective in North America? **Check the two that apply.**
    a. Concrete over steel deck over steel beams and steel columns framing system
    b. Panelized wood floor over open web steel truss over girder and columns
    c. Panelized wood floor over open web wood truss over girder and columns
    d. Panelized wood floor over purlins over girder and columns

21. Which of the following should not be used in direct contact with wood treated with waterborne preservatives containing copper?
    a. Aluminum
    b. Hot-dipped galvanized steel
    c. Copper
    d. Silicon bronze
    e. Stainless steel

22. Which of the following is the best definition of a horizontal exit?
    a. A horizontal exit is a two-hour separation, separating the building into two compartments.
    b. A horizontal exit is an exit on the same level.
    c. A horizontal exit is an exit enclosed by exit corridor on the same level.
    d. A horizontal exit is always required on every building. It is a basic form of exit.

23. Which of the following is the most true about travel distance? **Check the two that apply.**
    a. A travel distance is measured from the door of the most remote room, within a story.
    b. A travel distance is measured from the most remote point, within a story.
    c. When the path of exit access includes unenclosed stairway, the distance of travel on the stair shall be included in the travel distance measurement.
    d. When the path of exit access includes unenclosed stairway, the distance of travel on the stair shall not be included in the travel distance measurement.

24. In a building equipped with an automatic sprinkler system throughout, the separation distance of the exit doors or exit access doorways shall not be less than:
    a. One quarter of the length of the maximum overall diagonal dimension of the area served.
    b. One-third of the length of the maximum overall diagonal dimension of the area served.
    c. One-half of the length of the maximum overall diagonal dimension of the area served.
    d. Three quarters of the length of the maximum overall diagonal dimension of the area served.

25. Which of the following are the basic federal laws, which involve accessibility issues? **Check the two that apply.**
    a. Americans with Disabilities Act
    b. Fair Housing Act
    c. Fair Employment & Housing Act
    d. Unruh Civil Rights Act

26. An architect is designing an accessible counter and sinks in a public restroom. Which of the following is correct?
    a. The accessible counter has to be 2'-10" (864) from the finish floor.
    b. The tops of the rims of the sinks, on the accessible counter, have to be 2'-10" (864) from the finish floor.
    c. The accessible counter has to be 2'-10" (864), maximum, from the finish floor.
    d. The tops of the rims of the sinks, on the accessible counter, have to be 2'-10" (864), maximum, from the finish floor.

27. Which of the following statements are not true? **Check the two that apply.**
    a. All accessible counters have to have knee space below the counters.
    b. Accessible counters for workstations have to have knee space below the counters.
    c. Some accessible counters have to have knee space below the counters.
    d. Accessible transaction counters have to have knee space below the counters on the customer side.

28. An architect is preparing the plan check package for a retail store. Which of the following statements are true? **Check the two that apply.**
    a. If the store sells fresh produce and meat, the architect probably needs to submit the plans to Health Department for plan check.
    b. If the store sells clothing, the architect probably needs to submit the plans to Health Department for plan check.
    c. If the store sells candies, the architect probably needs to submit the plans to Health Department for plan check.
    d. If the store sells candies, the architect probably does not need to submit the plans to Health Department for plan check.

29. According to *International Building Code* (IBC), interior adhered masonry veneers shall have a maximum weight of _____psf (or _____ kg/ m^2).

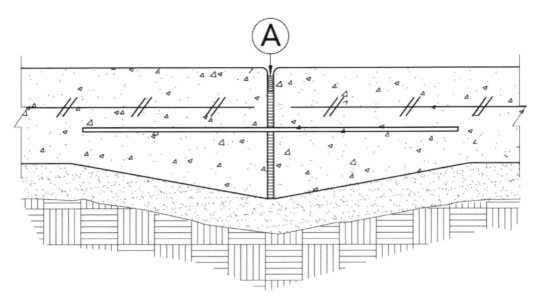

Figure 3.5   Concrete detail

30. What is noted as letter A on the previous image?
    a. Construction joint
    b. Control joint
    c. Concrete separation
    d. Concrete mark

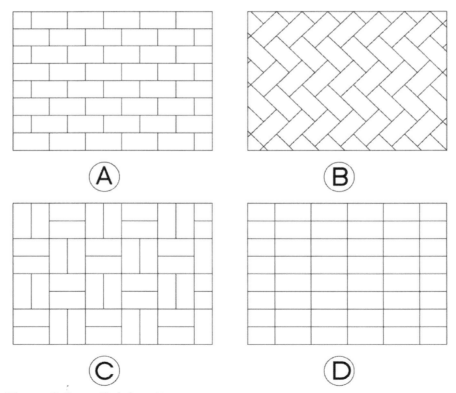

Figure 3.6   Brick pattern

31. Which pattern/letter on the previous image is stack bond?
    a. Pattern A
    b. Pattern B
    c. Pattern C
    d. Pattern D

32. Which pattern/letter on the previous image is likely to have the highest construction cost?
    a. Pattern A
    b. Pattern B
    c. Pattern C
    d. Pattern D

33. What is the purpose of the weep holes at the bottom of the CMU retaining walls?
    a. To vent the air from the bottom of the retaining wall
    b. To let small insects pass through the retaining walls to preserve biodiversity
    c. To drain water from the bottom of the retaining wall
    d. To provide room for the expansion of CMU

34. The term "PVC" is likely to appear in the specifications of which of the following?
    a. Built Up Roof
    b. Single Ply Roof
    c. Tile Roof
    d. Wood Shingle Roof

35. Which of the following statements are true? **Check the two that apply.**
    a. Single-glass windows typically have a higher R-value than double-glass windows.
    b. Single-glass windows typically have a lower R-value than double-glass windows.
    c. Single-glass windows typically have a higher U-value than double-glass windows
    d. Single-glass windows typically have a lower U-value than double-glass windows

36. For hydraulic elevators, the depth of the piston cylinder well is:
    a. equal to 1/2 the height of elevator travel
    b. equal to 1/3 the height of elevator travel
    c. equal to the height of elevator travel
    d. equal to twice the height of elevator travel

37. A client selects a site with underground methane gas to build a supermarket. Which of the following statements are true? **Check the two that apply.**
    a. The contractor should use special construction techniques to avoid explosion at the site.
    b. The contractor should have his employee use special masks at the site.
    c. The contractor can use underground-perforated pipes to collect the methane gas.
    d. The contractor can use automatic vent damper devices to alleviate the methane gas problem

38. Why is Radon gas not desirable in a building project?
    a. It has bad odor.
    b. It is poisonous.
    c. It is radioactive.
    d. It has too much moisture and is a source of mold problem.

39. An architect is working on a department store. The owner's prototype requires the panic hardware to have a 15-second delay to prevent theft of store merchandise. When the architect submits the plans to the fire department, the plan checker requires the architect to change to a panic hardware without any time delay function. What are the improper actions for the architect? **Check the two that apply.**
    a. Change to a panic hardware without any time delay function per the plan check corrections.
    b. Change to a panic hardware without any time delay function only at the accessible exit doors.
    c. Talk with the plan checker and find out if she can add a sign stating "Keep pushing, the door will open after 15 seconds" as an alternative solution.
    d. Change to a panic hardware without any time delay function per the plan check corrections to get plan check approval, and then tell the owner he can change back to the panic hardware with a 15-second delay after the building is completed.

40. Which of the following toilet partition finishes has the highest initial cost?
    a. Baked enamel
    b. Laminated plastic
    c. Porcelain enamel
    d. Powder shield
    e. Stainless steel
    f. Polly

41. Which of the following normally do not require panic hardware at the required exits? **Check the two that apply.**
    a. Museums
    b. Warehouses with no hazardous materials
    c. Post offices
    d. Restaurants

42. Per IBC, which of the following is correct regarding the panic hardware at the required exits?
    a. The actuating portion of the releasing device shall extend at least one-quarter of the door leaf width.
    b. The actuating portion of the releasing device shall extend at least one-half of the door leaf width.
    c. The actuating portion of the releasing device shall extend at least three-quarter of the door leaf width.
    d. The actuating portion of the releasing device shall extend the entire door leaf width.

43. All of the following will affect window selection except:
    a. Building orientation
    b. Location of the window
    c. The low initial cost of low-e glass
    d. The high initial cost of low-e glass

44. Where should the vapor barrier be installed?
    a. On the inside of the wall insulation
    b. On the outside of the wall insulation
    c. On the warm side of the walls
    d. On the cold side of the walls

45. Per 2902.1 of IBC, the minimum number of required water closets for Mercantile Occupancy Group is 1 per 500 occupants. An architect is designing a retail store with 32,000 square feet of gross area. All the store areas are on the ground level, and 5% of the areas are storage. The remaining areas are all retail spaces. Based on the following table, the minimum number of water closets required for Women's Restroom for this store is _____.

Table 3.1   Maximum Floor Area Allowances per Occupant

Mercantile	
Areas on other floors	60 gross
Basement and grade floor areas	30 gross
Storage, stock, shipping areas	300 gross

46. Architects often specify Type "X" Gypsum Wallboard for
    a. Low cost
    b. Recyclability
    c. Ease of installation
    d. Fire resistance

47. The ability of water to flow against gravity through concrete floor cracks is called
    a. Capillary action
    b. Seepage
    c. Saturation
    d. Leakage

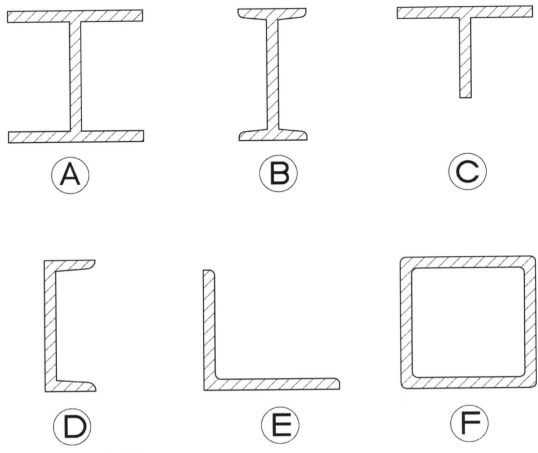

Figure 3.7   Steel shapes

48. Which of the previous images shows a W shape steel?
    a. A
    b. B
    c. C
    d. D

49. Right after the punch walk of a project, a HVAC worker falls down through the roof
    hatch opening and dies. The constructions plans do not show any safety railing around
    the roof hatch opening. Who is responsible for this accident?
    a. The owner
    b. The Architect
    c. The contractor
    d. The HVAC subcontractor

50. Per *International Mechanical Code* (IMC), if the height of a roof access ladder is over ___
    _____ feet, an intermediate landing is required.

51. Which of the following is not true about insulation? **Check the two that apply.**
    a. Batt insulation can be attached to the bottom of the roof deck.
    b. Batt insulation can be installed under the concrete slab.
    c. Rigid insulation can be installed under the concrete slab.
    d. Rigid insulation is typically attached to the bottom of the roof deck.

52. Which of the following has the least impact on the design of elevators?
    a. Accessibility
    b. Safety
    c. Number of passengers at peak hour
    d. Building's population

53. The nominal size of a standard brick in the US is _____, and the nominal size of a standard concrete masonry unit (CMU) in the US is _____. The actual size is usually about _____ smaller to allow for mortar joints.

54. An architect is trying to locate the main entrance of a CMU building, according to CMU block module, and use either the CMU full block or CMU half block to avoid cutting the CMU blocks. Which of the following dimension fits the CMU block module? **Check the two that apply.**
    a. 18'-4"
    b. 19'-4"
    c. 20'-0"
    d. 20'-4"

55. An architect is designing a federal building with an escalator between first and second floor. Which of the following is true? **Check the two that apply.**
    a. She needs to determine the vertical slope of the escalator.
    b. She needs to use dimensions to locate the work points for the escalator.
    c. She needs to draw and add full dimensions for the escalator so that the contractor can build the escalator accurately.
    d. She needs to coordinate with structural and electrical engineers

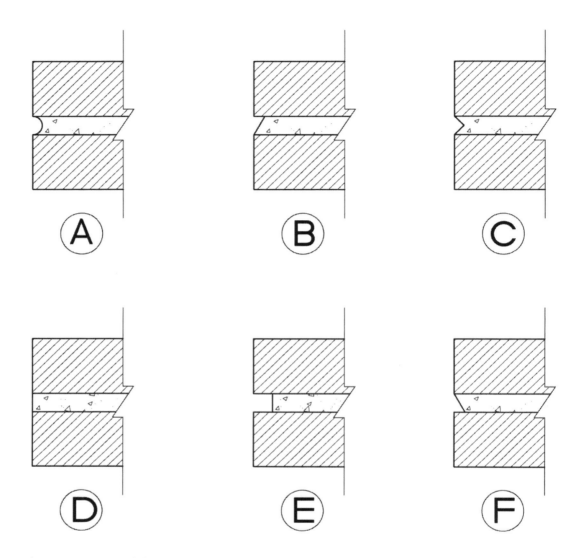

Figure 3.8   Brick joints

56. Which of the previous images shows a Weather Struck brick joint?
    a. A
    b. B
    c. C
    d. D

57. Strips made of white polyethylene are often added to the top and side of glass block partitions. The purpose of the strips is:
    a. To absorb extra moisture from the glass block partition.
    b. To protect the glass block partition in an earthquake.
    c. To provide room for glass block expansion.
    d. None of the above

58. The purpose of a sliptrack on the top of the full-height metal stud walls is:
    a. To protect the metal stud walls from wind forces.
    b. To protect the metal stud walls in an earthquake.
    c. To provide room for metal stud expansion.
    d. All of the above

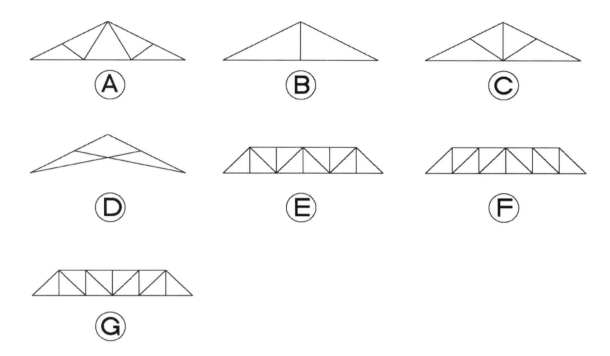

Figure 3.9   Truss types

59. Which of the previous images shows a fink truss?
    a. A
    b. B
    c. C
    d. D
    e. E
    f. F
    g. G

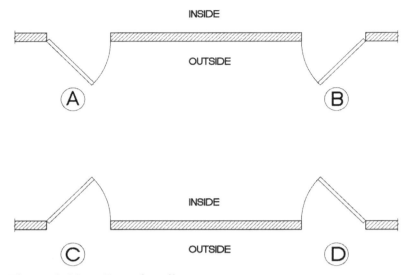

Figure 3.10   Door handing

60. Which of the previous images shows a right-hand door?
   a.   A
   b.   B
   c.   C
   d.   D

61. Which of the following are the federal laws that mandate accessibility to certain historic structures? **Check the three that apply.**
   a.   Americans with Disabilities Act
   b.   Fair Housing Act
   c.   Fair Employment & Housing Act
   d.   Unruh Civil Rights Act
   e.   Architectural Barriers Act
   f.   Section 504 of the Rehabilitation Act

62. Which of the following is the best solution to comply with accessibility requirements in a 2-story historic museum?
   a.   Install an elevator to provide access to the second floor
   b.   Install a lift to provide access to the second floor
   c.   Provide a refuge area next to the stair on the second floor
   d.   Provide equivalent exhibits on the first floor for handicapped people

63. Which of the following is not recommended when working on the entrance of a historic building?
   a.   No documentation on the new work
   b.   Stabilizing
   c.   Repairing
   d.   Replacement

64. If some of the exposed beams of an historic building have deflection, which of the following is recommended? **Check the two that apply.**
   a. Leaving them in place without doing anything
   b. Augmenting or upgrading
   c. Protecting and maintaining the beams and ensuring that structural members are free from insect infestation.
   d. Replacement of all the beams

65. Which of the following is not recommended when working on the preservation of the interior of a historic building? **Check the two that apply.**
   a. Installing protective coverings for wall coverings in the corridors
   b. Replacement of broken window glazing
   c. Using propane to remove paint
   d. Sandblasting of character-defining features

66. A new mechanical system is required to make a historic building functional. Which of the following is recommended?
   a. Place the mechanical unit in the attic and remove a substantial amount of building materials.
   b. Place the mechanical unit in an existing masonry unit enclosure and cut through the existing masonry walls.
   c. Place the mechanical unit on the ground and enclose it with hedges.
   d. Place the mechanical unit on the roof top and add additional beams and columns to support the unit

67. Which of the following procedure is recommended for historic building?
   a. Identifying, retaining, and preserving, stabilizing, protecting and maintaining, repairing, and replacement
   b. Identifying, retaining, and preserving, protecting and maintaining, repairing, replacement, and stabilizing
   c. Identifying, retaining, and preserving, repairing, replacement, stabilizing, protecting and maintaining
   d. Identifying, repairing, replacement, stabilizing, protecting and maintaining, retaining, and preserving

68. After code research, an architect discovers an exposed wood stair in a historic building has to be fire rated. Which of the following procedure is recommended? **Check the two that apply.**
   a. Enclosing the wood stair with fire-resistant sheathing required by codes
   b. Upgrading historic stairway so that it is not damaged or obscured
   c. Adding a new stairway or elevator to meet health and safety codes in a manner that preserves adjacent character-defining features and spaces
   d. Installing sensitively designed fire suppression systems, such as sprinkler systems that result in retention of historic features and finishes

69. The lead-based paint for a historic building start to peel, chip, craze, or otherwise comes loose. Which of the following procedure is recommended?
   a. Leave the paint in place and do nothing
   b. Use the same kind of lead-based paint to repaint the damaged area
   c. Remove the lead-paint throughout the building and apply a compatible primer and finish paint
   d. None of the above

70. According to *The Secretary of the Interior's Standards for the Treatment of Historic Properties*, which of the following treatments are appropriate for a building individually listed in the National Register? **Check the two that apply.**
   a. Preservation
   b. Restoration
   c. Rehabilitation
   d. Reconstruction

71. Which of the following is the most appropriate structural system for a lab building that is sensitive to vibration?
   a. Cast-in-place concrete beam-and-slab system
   b. Heavy timber construction with panelized floor
   c. Lightweight concrete over metal deck over steel joists
   d. 4" gypsum concrete topping slab over wood deck over wood joists

72. A contractor submits structural steel shop drawings for the architect's review. The architect forwards the structural steel shop drawings to the structural engineer. Both the architect and the structural engineer have reviewed and marked up the shop drawings. The structural engineer also stamps and signs the shop drawings. After the structural steel members are installed in the field, per the approved shop drawings, the contractor notices the structural columns are too long, and will NOT achieve the roof slope required by the roof plans. The contractor submits a change order of $22,000 and one extra week of construction time for adjusted the structural columns. What is a proper action for the architect?
   a. Approves the change order and back charges the cost to the structural engineer.
   b. Approves the change order and submits a claim to her professional liabilities insurance company and to the structural engineer's professional liabilities insurance company.
   c. Negotiates with the contractor to reduce the amount of change order, obtains the owner's approval and then approves the change order.
   d. Denies the change order.

73. The principal AIA documents A201 family includes:
    a. Some documents in the A series, like agreement between owner and contractor
    b. Some documents in the A series, like agreement between owner and contractor and agreement between contractor and subcontractor
    c. Some documents in the A series, like agreement between owner and contractor and agreement between contractor and subcontractor, and some documents in the B series, like agreement between owner and architect
    d. Some documents in the A series, like agreement between owner and contractor and agreement between contractor and subcontractor, some documents in the B series, like agreement between owner and architect, and some documents in the C series, like agreement between architect and consultant

74. Which of the following is appropriate for extinguishing a fire in a cell phone equipment room? **Check the two that apply.**
    a. Dry ice
    b. Water
    c. Dry chemicals
    d. A combination of water, carbon dioxide and dry chemicals

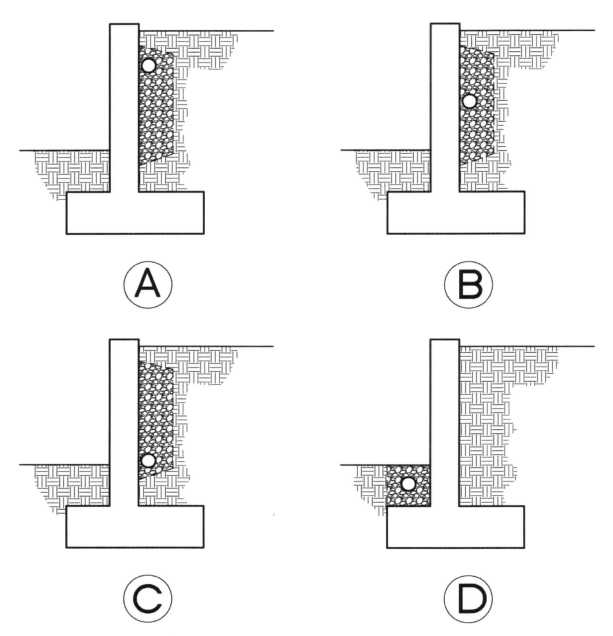

Figure 3.11   The perforated pipe location

75. Which of the previous images shows the correct location of a perforated pipe?
    a. A
    b. B
    c. C
    d. D

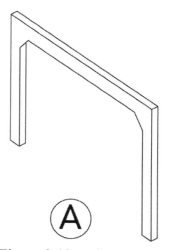

Figure 3.12   Structural element

76. Which structural element does the previous image show?
    a. Column and beam
    b. Rigid frame
    c. Space frame
    d. Prefabricated column and beam

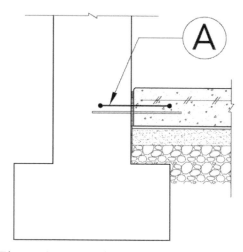

Figure 3.13   The correct term for the preformed synthetic rubber

77. Which of the following is the correct term for the preformed synthetic rubber labeled as
    letter A in previous image (**Note:** some other elements NOT shown for clarity)?
    a. Waterstop
    b. Moisture barrier
    c. Control joint
    d. Isolation joint

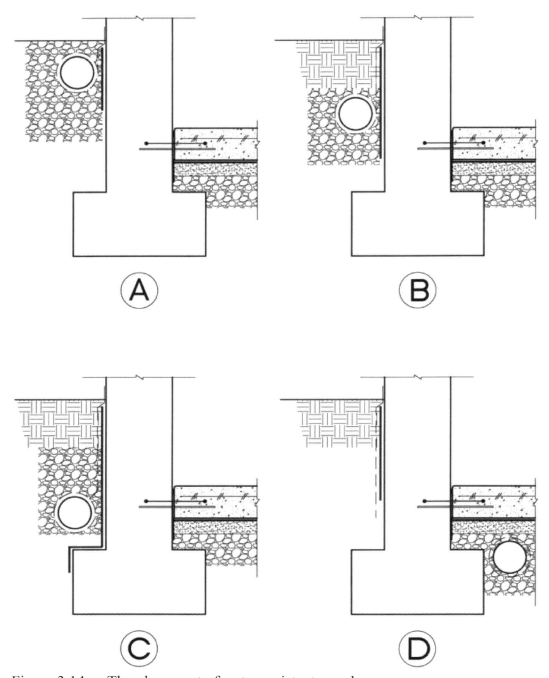

Figure 3.14   The placement of water resistant membrane

78. Which of the previous images shows the correct placement of water resistant membrane (shown as heavy solid lines. **Note:** some other elements are NOT shown for clarity)?
    a. A
    b. B
    c. C
    d. D

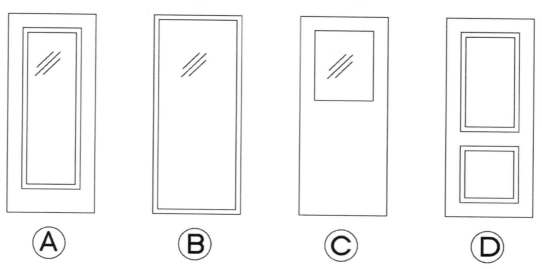

Figure 3.15   The door types by design

79. Which of the previous images shows a "Sash" type door design?
    a. A
    b. B
    c. C
    d. D

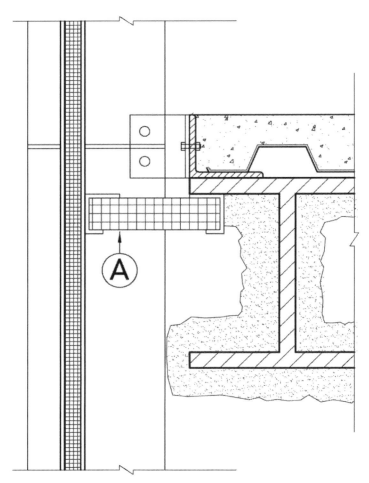

Figure 3.16   Sectional detail

80. Which of the following is the correct term for A in previous image?
    a. Beam plate
    b. Sliptrack
    c. Firestopping
    d. Steel support

81. In a fast track, one-story project, which of the following is most likely a critical path item?
    a. Roof framing system design
    b. HVAC system design
    c. Interior finish selection
    d. Door selection

82. Which of the following situations shall utilize grade beams? **Check the two that apply.**
    a. At site with expansive soils
    b. At storefront with no solid walls
    c. At site with underground rock
    d. At site in an inland area

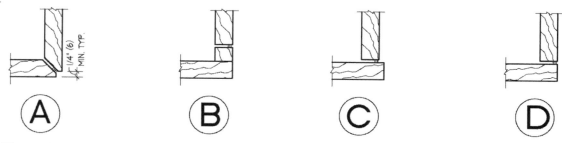

Figure 3.17   Veneer stone corner joints

83. Which of the following is the correct term for previous image A?
    a. Corner "L"
    b. Quirk miter
    c. Slip corner
    d. Butt joint

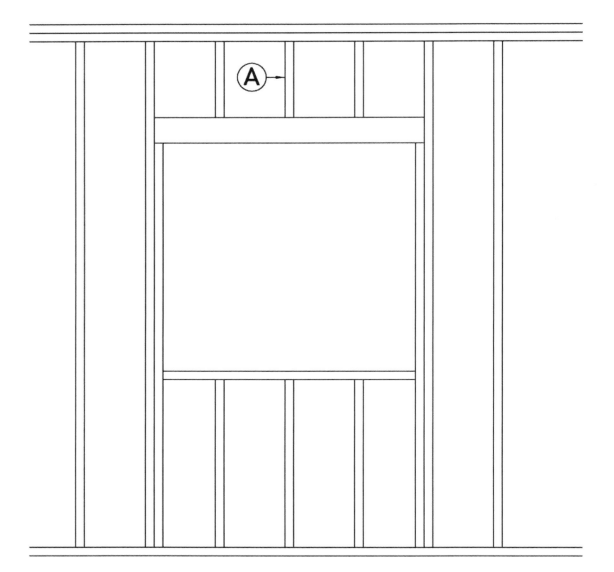

Figure 3.18   Framing detail

84. Which of the following is the correct term for A in previous image?
    a. Cripple studs
    b. King posts
    c. Transom studs
    d. Top studs

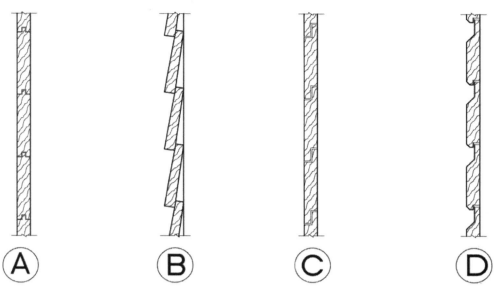

Figure 3.19   Wood siding

85. Which of the following is the correct term for the wood siding A in previous image?
    a. Bevel
    b. Shiplap
    c. Square edge tongue and groove
    d. Channel rustic

B. Mock Exam: BDCS Accessibility/Ramp Vignette

1. Directions and Code

Directions and **Code** are the same as the accessibility/ramp sample vignette in the official NCARB ARE BDCS exam guide. The NCARB **Directions** and **Code** has been very consistent through various versions of ARE.

You can download the official exam guide and practice program for BDCS division at the following link:
http://www.ncarb.org/en/ARE/Preparing-for-the-ARE.aspx

2. Program
1) You need to connect a new building addition to the existing office building that has a different floor elevation (Figure 3.20).
2) Use a ramp and a separate stair to connect the new lobby to the corridor of the existing building.
3) Add wall(s) and door(s) on the existing building to separate the lobby from the existing corridor.
4) The ramp and stair must also meet the following requirements:
   - The ramp and stair cannot encroach on the existing building
   - Show elevations on ALL landings

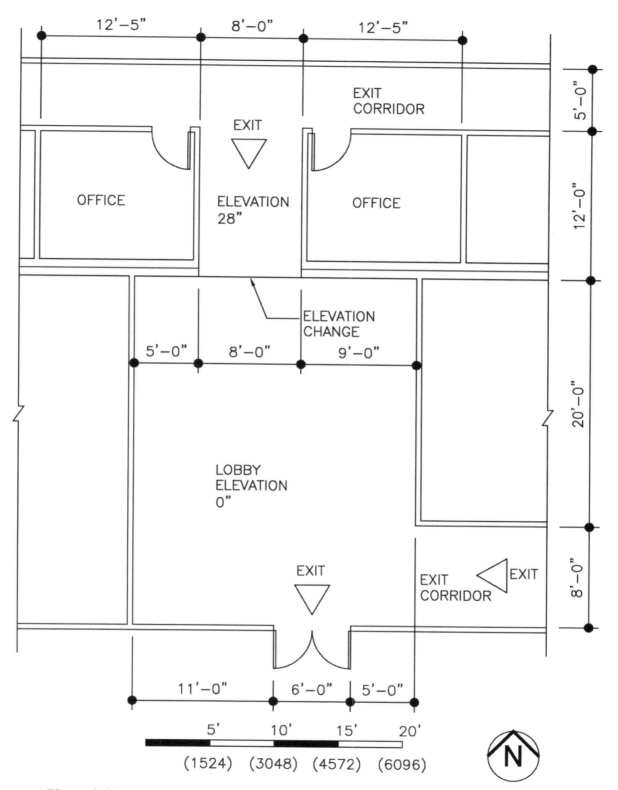

Figure 3.20   Ramp and stair vignette

C. Mock Exam: BDCS Stair Design Vignette

1. Directions and Code

Directions and **Code** are the same as the stair design sample vignette in the official NCARB ARE BDCS exam guide. The NCARB **Directions** and **Code** has been very consistent through various versions of ARE.

You can download the official exam guide and practice program for BDCS division at the following link:
http://www.ncarb.org/en/ARE/Preparing-for-the-ARE.aspx

2. Program

1) In an existing stair well, develop a new exit stair for a two-story office building to meet the accessibility requirements and new occupants load (Figure 3.32).
2) The second floor needs to have an area of refuge.
3) There is an existing second stairway near the exiting main building entrance at another location.
4) The stairway shall serve as a mean of egress for all three building levels discharge through the ground level exit door to the sidewalk, a public way at grade.
5) The stair shall provide a continuous path from the second floor to the first floor exit and includes a landing at the intermediate level.
6) The number of exits and total occupant loads for each level are:

Building Level	Number of Exits	Total Occupant Loads
Ground Floor	3	350
Storage	1	8
Second Floor	2	160

7) The stairs are pre-cast concrete with the following dimensions:
- Landings: 12" (305) deep between the surface and the landing soffit
- Stair flight/stringer: 12" (305) deep between the stringer soffit and the stair nosing measure along a line perpendicular to the soffit

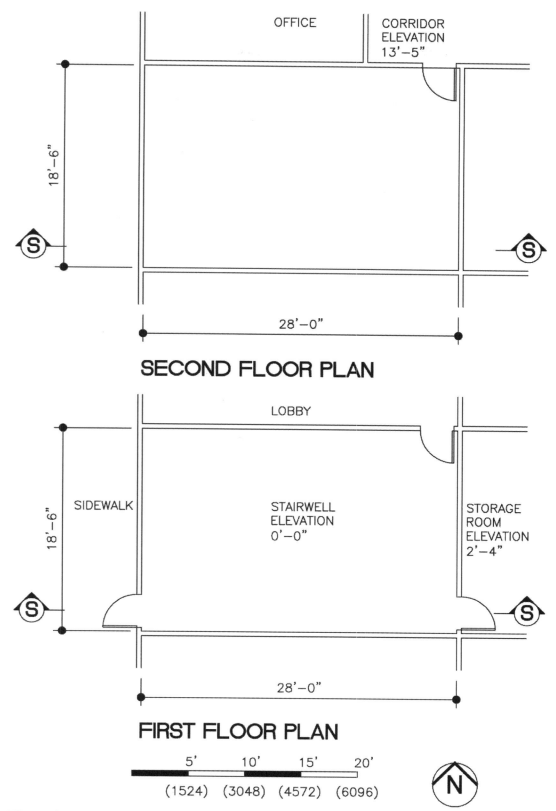

Figure 3.21   Stair vignette plans

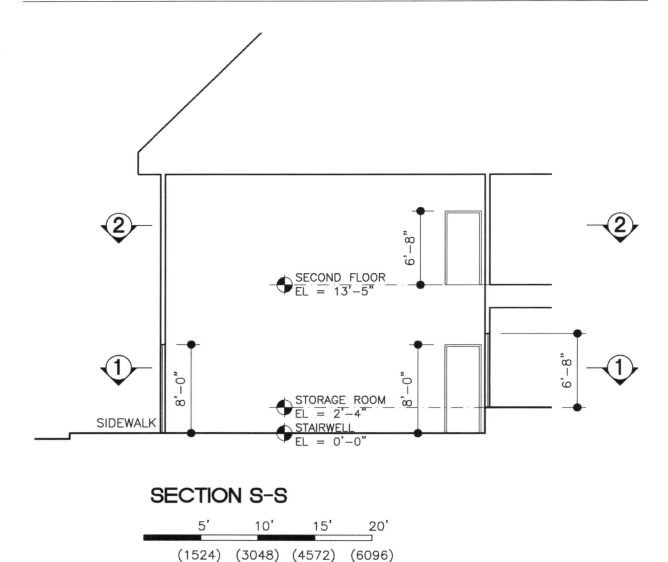

SECTION S-S

5'   10'   15'   20'

(1524) (3048) (4572) (6096)

Figure 3.22   Stair vignette section

D. Mock Exam: BDCS Roof Plan Vignette

1. Directions

Directions are the same as the roof plan sample vignette in the official NCARB ARE BDCS exam guide. The NCARB **Directions** have been very consistent through various versions of ARE.

You can download the official exam guide and practice program for BDCS division at the following link:
http://www.ncarb.org/en/ARE/Preparing-for-the-ARE.aspx

2. Program

You are designing an office building, and need to design a roof plan per the following requirements:

1) The building has two volumes, one low and one high. Each volume has its own roof height and slope requirements (Figure 3.33).
2) The site is in a temperate climate, and has moderate annual rainfall.
3) You can only use roof slope, downspout, and roof gutter to remove rainwater.
4) Downspouts shall not conflict with any window, clerestory window and door.
5) Rainwater shall not discharge from any gutter or roof directly onto the ground, or from the edge of an upper roof directly onto a lower roof.
6) Finish floor elevation is 0'-0". Minimum ceiling height is 8'-0" (2438).
7) All roof areas shall have positive drainage.
8) The slope for the Multi-purpose Room should be between 6:12 and 12:12.
   - The total thickness for the roof and structural assembly is 18" (457).
9) The slope for the remaining spaces should be between 2:12 and 5:12.
   - The total thickness for the roof and structural assembly is 18" (457).
10) The Multi-purpose Room has a 30-inch-high (762-high) continuous horizontal clerestory window in the existing east wall.
    - The clerestory sill is included in the over height dimension.
11) All rooms shall have natural light via a skylight, windows, or clerestory window.
    - You can only use skylights for rooms without windows, or clerestory window.
    - Storage rooms, closets, and halls do NOT require skylights.
12) You need to show flashing for all roof/wall surfaces intersection, including chimneys.
13) Gutters, skylights, exhaust fan vents, plumbing vent stacks, and HVAC condensing units are self-flashing, and do not require additional crickets or flashing.
14) You need to place the HVAC condensing unit on a roof with a slope of 5:12 or less.
    - Do not place in front of the clerestory window.
    - Maintain a minimum of 3'-0" (914) clearance from all roof edges.
15) Provide (1) exhaust fan for each toilet room.
16) Provide vent stacks through roof where required to vent plumbing fixtures.

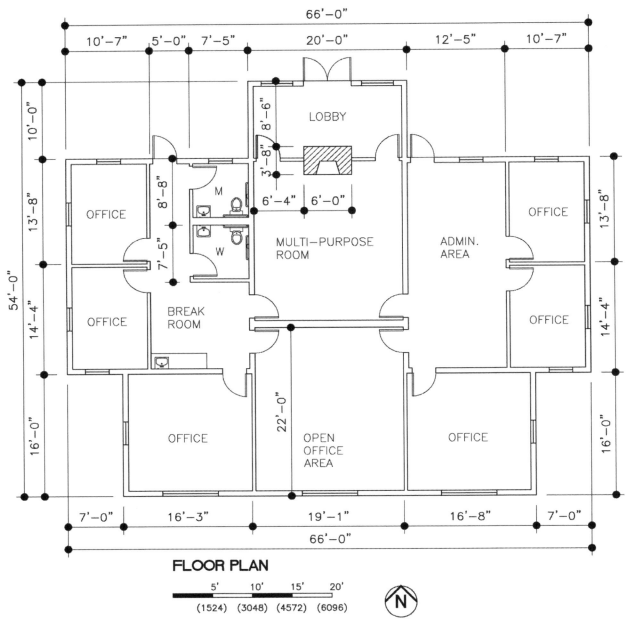

FLOOR PLAN

Figure 3.23   Floor plan for roof plan vignette

Chapter Four

ARE Mock Exam Solution for
Building Design and Construction Systems (BDCS) Division

A. Mock Exam Answers and explanation: BDCS Multiple-Choice (MC) Section

Note: If you answer 60% of the questions correctly, you pass the MC Section of the exam.

1. Answer: a and d
    Stainless steel nails and terne-coated stainless steel can be specified to affix shingles onto a roof structure. **Terne** is an alloy coating that used to be made from 20% tin and 80% lead. Currently, lead is not desirable in the construction industry. It has been replaced with the metal zinc and terne is made of 50% tin and 50% zinc. Terne metal must be painted. Well-maintained painted terne metal can last 90 years or more.

    Iron nails are not resistant to corrosion, and are NOT a good choice.

    Pneumatic staple guns can result in crushing the wood fibers, in shooting staples through the shingles, or in cracking the shingle. They are not a good choice either.

2. Answer: b and d
    Lead-based paint and asbestos have been banned in new construction since the late 1970s because they are hazardous materials.

    Copper nails can be used. Zinc-based paint is just a **distractor**.

3. Answer: c
    Red-cedar shingles should not have direct contact with copper nails because a chemical reaction between the copper and the wood will reduce the life of the roof. Hot-dipped, aluminum, zinc-coated, or stainless steel nails should be used.

    4-ply roofing is built-up roofing. TPO roofing is one kind of single-ply roofing. They both can be used with copper nails.

    Fiberglass roofing can also be used with copper nails.

4. Answer: d
    A cylinder lock is shown on the image. There are four major categories of locks:
    *   Mortise lock (fits into **mortise** on the door **edge**)
    *   Unit lock (fits into door **cutout**)
    *   Integral lock (A **combination** of cylinder lock and mortise lock)

- Cylinder lock (The most common and cost-effective type, used in many residential buildings. It fits into lock stile of the door and the **drilled holes**)

See *Building Construction Illustrated* for more information. You need to be able to identify the different kinds of locks when looking at the sketches or images.

5. Answer: b and d
   Please note that we are looking for the **incorrect** answers.
   The following statements are incorrect, and therefore the correct answers:
   - Control joints also serve as isolation or construction joints. (Control joints do NOT separate the slab completely, and their depths are normally ¼ of the slab thickness. Control joints can NOT serve as construction joints)
   - Control joints normally run from the top of slab to the bottom of the slab. (Control joints normally do **NOT** run from the top of slab to the bottom of the slab. their depths are normally ¼ of the slab thickness)

   The following statements are correct, and therefore the incorrect answers:
   - Construction joints also serve as isolation or control joints.
   - Construction joints normally run from the top of slab to the bottom of the slab.

   Control joints can be formed by saw cutting the concrete or by placing a 1/8" premolded strip insert when concrete is cast. They allow the concrete to crack along the predetermined lines.

6. Answer: a
   The term "fine sand float finish" refers to plastering.

   See the following link:
   http://www.tsib.org/plastertextures.shtml

   You need to read the information at the link above, and become familiar with various kinds of plaster finishes. This will not only help you pass ARE, but also help you in real architectural practice. Plaster finishes are normally noted on building exterior elevations.

   Paving, painting, and concrete are all **distractors**.

7. Answer: b
   The term "VOC" refers to Volatile Organic Compound. The word "Volatile" describes a liquid that evaporates at room temperature, and the word "organic" means a compound that contains carbon.

   Valid Organic Compound, Volatile Original Compound, and Volatile Original Composite are all **distractors**.

8. Answer: b
   A pivoting window is shown on the image.

The major window types by movement include:
a. Sliding
b. Pivoting
c. Casement
d. Fixed
e. Awning/Hopper
f. Double-hung
g. Jalousie

You need to be able to identify the window types by reviewing the images or line drawings.

See *Building Construction Illustrated* for more information:

9. Answer: c
    Letter C in the figure indicates a mullion.

    Letter A in the figure indicates a jamb.

    Letter B in the figure indicates a head.

    Letter D in the figure indicates a sill.

10. Answer: d
    Plastering is the most cost-effective finish for exterior walls.

11. Answer: a
    According to *International Building Code* (IBC), the handrails shall extend at least 12" (305) beyond the top riser and continue to slope for the depth of one tread beyond the bottom.

    See Section 1012.5 of *International Building Code* (IBC).

    See following link for the FREE IBC code sections citations:
    http://publicecodes.citation.com/icod/ibc/2006f2/icod_ibc_2006f2_10_sec012_par004.ht m?bu=IC-P-2006-000001&bu2=IC-P-2006-000019

12. Answer:
    According to *International Building Code* (IBC), stairways shall have a minimum headroom clearance of 80" (2032) measured vertically from a line connecting the edge of the nosing.

    See Section 1009.2 of *International Building Code* (IBC). See following link for the FREE IBC code sections citations:
    http://publicecodes.citation.com/icod/ibc/2006f2/icod_ibc_2006f2_10_sec009_par001.ht m?bu=IC-P-2006-000001&bu2=IC-P-2006-000019

13. Answer:
According to *International Building Code* (IBC), stair riser heights shall be 7" (178) maximum and 4" (102) minimum. Stair tread depths shall be 11" (279) minimum.

See Section 1009.3 of *International Building Code* (IBC).

See following link for the FREE IBC code sections citations:
http://publicecodes.citation.com/icod/ibc/2006f2/icod_ibc_2006f2_10_sec009_par002.htm?bu=IC-P-2006-000001&bu2=IC-P-2006-000019

14. Answer: c
The term "Elastomeric - Modified Acrylic" is most likely to be found in the specifications for **painting**. Elastomeric means elastic. Acrylic means a paint containing acrylic resin

15. Answer: b and c
The best places to find the information for the 8' high FRP are room finish schedules and interior elevations because the 8' high FRP are interior room finish for the food prep areas.

16. Answer: c
She should add a cricket on the **high** side of the roof next to the chimney to drain and divert the rainwater away from the chimney.

Rainwater flows from high to low, so there is no point of adding a cricket on the **low** side of the roof next to the chimney.

A cricket should ALWAYS be placed on the **high** side of the roof next to the chimney, HVAC units, and condensing units to divert the rainwater.

17. Answer: a, b, c and f
In a building that uses the Insulated Concrete Forms (ICF) system, the forms are left in place permanently for the following reasons:
- Thermal insulation
- Acoustic insulation
- Space to run plumbing pipes and electrical conduits. The form material on either side of the walls can easily accommodate plumbing and electrical installations.
- Backing for stucco, brick, or other siding on the exterior and gypsum boards on the interior

Aesthetic effect and construction cost are possible answers, but they are not as good as the other answers.

18. Answer: b
A **drip edge** is noted as letter A on the image.

Flashings are used at the intersections of walls, roof, chimney, etc.
Drip edges are used at the edge of parapets or eaves to drip rainwater.
Copings are used to cover the top of parapets and may extend to the edge of the parapet.
Sidings are the finishes for exterior walls.

*Note: The most important tip for quality control of a project is "**Don't leak and don't fall.**" For the "don't fall" part, you need to pick a good structural engineer, and you need to do a good job to coordinate with him. For the "**Don't leak**" part, you need to make sure ALL your exterior details need to work and keep the water and moisture out of a building. A **drip edge** is an important detail to keep water out of a building.*

19. Answer: b, d and e
    If a set of building plans has to be reviewed by the Health Department, the following are **likely** to be acceptable by the Health Department as floor finish for the Janitor's room because they can be cleaned easily:
    - Concrete with slim foot base
    - Sheet vinyl flooring with cove base
    - Ceramic tile floor with cove base

    The following are **unlikely** to be acceptable by the Health Department as floor finish for the Janitor's room because they cannot be cleaned easily, or are NOT tolerant of water and moisture:
    - Carpet with wood base
    - VCT flooring with cove base (VCT stands for vinyl composition tile)
    - Smooth wood floor with cove base

20. Answer: b and c
    The following is likely to be most cost-effective in North America:
    - Panelized wood floor over open web steel truss over girder and columns
    - Panelized wood floor over open web wood truss over girder and columns

    The following is unlikely to be most cost-effective in North America:
    - Concrete over steel deck over steel beams and steel columns framing system
    - Panelized wood floor over purlins over girder and columns

    In North America, concrete over steel deck is typically more expensive than panelized wood floor. Panelized wood floor over purlins is typically more expensive than panelized wood floor over open web wood truss or panelized wood floor over open web steel truss.

21 Answer: a
    Waterborne preservatives can increase the risk of corrosion when metals contact treated wood used in wet locations.

Aluminum should not be used in direct contact with wood treated with waterborne preservatives containing copper. On the other hand, hot-dipped galvanized steel, copper, silicon bronze, or stainless steel can be used in direct contact with wood treated with waterborne preservatives containing copper

22 Answer: a

A horizontal exit is basically a **two-hour separation**, separating the building into two compartments, A and B. This two-hour separation needs to extend from exterior wall to exterior wall and separate the two parts of building completely. **When people from compartment A enter compartment B, they "exited" from compartment A**, and vice versa. Horizontal exits are extremely valuable for assembly occupancies like casino or convention spaces or a large restaurant on the top story of a high-rise building because they can avoid the use of huge exit stairs, and save spaces.

A horizontal exit is an exit on the same level, but its most important feature is the two-hour separation.

A horizontal exit is NOT an exit enclosed by exit corridor on the same level.

A horizontal exit is NOT always required on every building. It is NOT a basic form of exit.

23 Answer: b and c
The following is true about travel distance:
- A travel distance is measured from the most remote point within a story.
- When the path of exit access includes unenclosed stairway, the distance of travel on the stair shall be included in the travel distance measurement.

The following is not true about travel distance:
- A travel distance is measured from the door of the most remote room within a story.
- When the path of exit access includes unenclosed stairway, the distance of travel on the stair shall not be included in the travel distance measurement.

See Section 1016.1 of *International Building Code* (IBC).

See following link for the FREE IBC code sections citations:
http://publiccodes.citation.com/icod/ibc/2006f2/icod_ibc_2006f2_10_sec016.htm?bu2=undefined

24. Answer: b
In a building equipped with an **automatic sprinkler** system throughout, the separation distance of the exit doors or exit access doorways shall not be less than **one-third** of the length of the maximum overall diagonal dimension of the area served.

If the building is **NOT** equipped throughout with an automatic sprinkler system, the separation distance of the exit doors or exit access doorways shall not be less than **one-half** of the length of the maximum overall diagonal dimension of the area served.

See Section 101.2, exception #2 of *International Building Code* (IBC).

See following link for the FREE IBC code sections citations: http://publiccodes.citation.com/icod/ibc/2006f2/icod_ibc_2006f2_10_sec015_par004.ht m?bu2=undefined

25. Answer: a and b
    The following are the basic federal laws involve accessibility issues:
    * Americans with Disabilities Act
    * Fair Housing Act

    The following are the basic California laws involve accessibility issues:
    * Fair Employment & Housing Act
    * Unruh Civil Rights Act
    * Disabled Persons Act

26. Answer: d
    An architect is designing an accessible counter and the sinks in a public restroom. The tops of the rims of the sinks on the accessible counter have to be 2'-10" (864) maximum from the finish floor. It can be a little lower, and does not have to be 2'-10" (864) maximum from the finish floor.

    See more information at the following link: http://www.access-board.gov/adaag/html/figures/fig31.html

27. Answer: a and d
    Pay attention to the word "not." We are looking for the statements that are not true.

    The following statements are true, and therefore the incorrect answers:
    * Accessible counters for workstations have to have knee space below the counters.
    * Some accessible counters have to have knee space below the counters.

    The following statements are untrue, and therefore the correct answers:
    * All accessible counters have to have knee space below the counters. (NOT all accessible counters have to have knee space below the counters. For example, the accessible transaction counters in a store do <u>NOT</u> have to have knee space below the counters on the customer side.)
    * Accessible transaction counters have to have knee space below the counters on the customer side. (Accessible counters for transaction stations do <u>NOT</u> have to have knee space below the counters on the customer side.)

28. Answer: a and c

An architect is preparing the plan check package for a retail store. If the store sells any kinds of food including fresh produce, meat and candies (pre-package food), the architect probably needs to submit the plans to Health Department for plan check.

The following statements are true:
- If the store sells fresh produce and meat, the architect probably needs to submit the plans to Health Department for plan check.
- If the store sells candies, the architect probably needs to submit the plans to Health Department for plan check.

The following statements are not true:

- If the store sells clothing, the architect probably needs to submit the plans to Health Department for plan check. (clothing is not food, Health Department is typically not interested in checking plans for stores that do not sell foods)
- If the store sells candies, the architect probably does not need to submit the plans to Health Department for plan check.

29. Answer:

According to *International Building Code* (IBC), interior adhered masonry veneers shall have a maximum weight of **20** psf (or 97.6 kg/m^2).

Note: Exterior veneers shall have a maximum weight of 15 psf (or 73.2 kg/m^2).

These are important numbers to remember. It will help you in selecting the proper masonry veneers in architectural practice. Please note masonry veneers include both **stone** veneers and brick veneers.

There are two major categories of masonry veneers based on techniques of installation:
- Adhered masonry veneers (They are easy to install and more cost effective than Anchored masonry veneers)
- Anchored masonry veneers (They require corrosion-resistant fastenings, metal ties, etc.)

See Section 1405.9.1 of *International Building Code* (IBC).

See following link for the FREE IBC code sections citations:
http://publicecodes.citation.com/icod/ibc/2006f2/icod_ibc_2006f2_14_sec005_par013.htm?bu=IC-P-2006-000001&bu2=IC-P-2006-000019

The online IBC has a typo:

"Interior adhered masonry veneers shall have a maximum weight of 20 psf (0.958 kg/m^2)..."

2$\underline{0}$ psf (**0.958** kg/ m^2) is a typo, and it should be 20 psf (**97.6** kg/ m^2).

Here are my calculations:

1 pound = 0.4536 kilogram

1 sf = 0.0929 m^2

2$\underline{0}$ psf = (20 x 0.4536 kilogram) / 0.0929 m^2 = 97.6 kg/ m^2

30. Answer: a
    A construction joint is noted as letter A on the image.
    - Construction joints also serve as isolation or control joints.
    - Construction joints normally run from the top of slab to the bottom of the slab.
    - Control joints do NOT separate the slab completely, and their depths are normally ¼ of the slab thickness. Control joints can NOT serve as construction joints.
    - Concrete separation and concrete mark are just distractors.

31. Answer: d
    Pattern D on the previous image is stack bond.

    The following are names for all the patterns:
    a. Pattern A: Running bond
    b. Pattern B: Herringbone
    c. Pattern C: Basketweave
    d. Pattern D: Stack bond

32. Answer: b
    The Herringbone pattern/letter on the previous image is likely to have the highest construction cost because it is the most complicated pattern and takes the most labor to build.

33. Answer: c
    The purpose of the weep holes at the bottom of the CMU retaining walls is to drain water from the bottom of the retaining wall. All other answers are distracters.

34. Answer: b
    The term "PVC" is likely to appear in the specifications of Single Ply Roof.

    There are two major categories of membrane roofing systems:
    - Bituminous Systems (the traditional systems)
    - Single Ply Roof Systems (the newer systems)

    They can be further divided as following:

Built Up Roof is one subcategory of Bituminous Systems. Bituminous Systems includes three basic subcategories:
- Bituminous Systems - BUR (Built Up)
- Bituminous Systems - APP Modified Bitumen
- Bituminous Systems - SBS Modified Bitumen

Single Ply Roofing Systems includes several basic subcategories:
- Single Ply - TPO (an ethylene propylene rubber)
- Single Ply - PVC (a thermoplastic material)
- Single Ply – EPDM (an elastomeric material)
- Single Ply – CSPE (a synthetic rubber)
- Single Ply – Neoprene (a synthetic rubber)
- Single Ply – Polymer-modified Bitumen (a composite material)

You need to be familiar with these terms.

Tile Roofing and Wood Shingle Roofing are distractors.

35. Answer: b and c

The following statements are true:
- Single-glass windows typically have a lower R-value than double-glass windows.
- Single-glass windows typically have a higher U-value than double-glass windows

R-value indicates a material's ability to resist heat flow. Higher R-value indicates the better ability to resist heat flow.

U-value indicates a material's ability to transfer heat. Higher U-value indicates the better ability to transfer heat.

36. Answer: c

For hydraulic elevators, the depth of the piston cylinder well is equal to the height of elevator travel.

A hydraulic elevator does NOT need a penthouse, but it has lower speed and is limited to six-story height because of the piston length.

37. Answer: a and c

For a site with underground methane gas, the following statements are true:
- The contractor should use special construction techniques to avoid explosion at the site. (Methane is the major component of natural gas. Methane gas is highly combustible. Any normal construction techniques that can generate sparks or fire are prohibited for the danger of job site fire or explosion.)
- The contractor can use underground-perforated pipes to collect the methane gas. (The perforated pipes can collect the methane gas, and then vent the methane gas at a certain height above roof.)

The following statements are not true:
- The contractor should have his employee use special masks at the site. (The contractor does NOT need to have his employee use special masks at the site.)
- The contractor can use automatic vent damper devices to alleviate the methane gas problem. (Automatic vent damper devices are used to regulate airflow of the HVAC system. They have nothing to do with methane gas and can NOT alleviate the methane gas problem.)

38. Answer: c

Radon gas is not desirable in a building project because it is radioactive and is considered a health hazard.

Radon is atomic number 86, a chemical element with symbol Rn. It is an odorless, colorless, radioactive, tasteless noble gas, occurring naturally as the decay product of uranium (atomic number 92). It is considered a health hazard due to its radioactivity.

The following statements are not true:
- It has bad odor.
- It is poisonous.
- It has too much moisture and is a source of mold problem.

39. Answer: b and d

Please note we are looking for **improper** actions for the architect.

The following statements are improper actions for the architect and therefore the correct answer:
- Change to a panic hardware without any time delay function only at the accessible exit doors. (This does not work because the plan check corrections apply to ALL exit doors and the related panic hardware.)
- Change to a panic hardware without any time delay function per the plan check corrections to get plan check approval, and then tell the owner he can change back to the panic hardware with a 15-second delay after the building is completed. (This is unethical and against the NCARB **Rules of Conduct**.)

   Rules of Conduct is Available as a FREE PDF file at (Skimming through it should be adequate):
   http://www.ncarb.org/

The following statements are proper actions for the architect and therefore the incorrect answer:
- Change to a panic hardware without any time delay function per the plan check corrections.
- Talk with the plan checker and find out if she can add a sign stating "Keep pushing, the door will open after 15 seconds" as an alternative solution.

40. Answer: e
    Stainless steel toilet partition finishes has the highest initial cost.
    Baked enamel has the lowest initial cost. Powder shield is powder coated baked enamel.
    Polly is solid plastic HDPE. Porcelain enamel has a low initial cost also.

41. Answer: b and c
    Please pay attention to the word "not" in the original question.
    The following normally do <u>not</u> require panic hardware at the required exits, and therefore
    are the correct answers:
    - Warehouses with no hazardous materials
    - Post offices

    The following normally require panic hardware at the required exits, and therefore the
    incorrect answers:
    - Museums
    - Restaurants

    Per Section 1008.1.9 of IBC, each door in a means of egress from a Group A or E
    occupancy having an occupant load of 50 or more and any Group H occupancy shall not
    be provided with a latch or lock unless it is <u>panic hardware</u> or fire exit hardware.

    Per Section 303 of IBC, museums and restaurants belong to Assembly Group A, and
    require panic hardware at the required exits.

    Per Section 304 of IBC, post offices belong to Business Group B, and do <u>not</u> require
    panic hardware at the required exits.

    Per Section 311 of IBC, warehouses with no hazardous materials belong to Storage
    Group S, and do <u>not</u> require panic hardware at the required exits.

    See following links for the FREE IBC code sections citations:
    http://publicecodes.citation.com/icod/ibc/2006f2/icod_ibc_2006f2_10_sec008_par024.ht
    m?bu=IC-P-2006-000001&bu2=IC-P-2006-000019

    http://publicecodes.citation.com/icod/ibc/2006f2/icod_ibc_2006f2_3_section.htm?bu=IC-
    P-2006-000001&bu2=IC-P-2006-000019

42. Answer: b
    Per IBC, the following is correct regarding the panic hardware at the required exits:
    - The actuating portion of the releasing device shall extend at least one-half of the door
      leaf width.

    See Section 1008.1.9 of IBC at the following link:
    http://publicecodes.citation.com/icod/ibc/2006f2/icod_ibc_2006f2_10_sec008_par024.ht
    m?bu=IC-P-2006-000001&bu2=IC-P-2006-000019

43. Answer: c
Pay attention to the word "except."

The low-e glass has a visible light transmission value, and a low heat transfer coefficient. The low-e glass has a high initial cost, but is energy efficient, and can save the owner money over a long-term.

All of the following will affect window selection, and are therefore the <u>incorrect</u> answers:
- Building orientation
- Location of the window
- The high initial cost of low-e glass

44. Answer: c
Warm air and/or moisture would create condensation on the vapor barrier.

The vapor barrier should be installed on the **warm** side of the walls to prevent condensation from being absorbed into the insulation. In most cases, the warm side of the walls occurs on the **inside** of the wall insulation, but for an air-conditioned building in hot and humid climate, the warm side of the wall occurs on the **outside** of the wall insulation.

45. Answer:
Based on the table, the minimum number of water closets required for Women's Restroom for this store is **2**.

The following is detailed process of the calculations:

Total gross square footage of the building:	32,000 s.f.
Gross square footage for <u>female</u> occupants:	32,000 s.f./2 = 16,000 s.f.
5% of the areas are storage:	5% x 16,000 s.f. = 800 s.f.
The number of the female occupants for storage area: (See Table 3.1)	800 s.f./300 = 2.67
The remaining retail spaces:	16,000 s.f.- 800 s.f. =15,200 s.f.
The number of the female occupants for retail spaces:	15,200 s.f./30 = 506.67
Total number of the <u>female</u> occupants:	2.67 + 506.67 = 509.34 or about 509

Per 2902.1 of IBC, the minimum number of required water closets for Mercantile Occupancy Group is 1 per 500 occupants.

Therefore, the minimum number of water closets required for Women's Restroom for this store is 509/500 = 1.018 or 2.

Note: We always round up to the next whole number when calculating the required minimum number of water closets, unless the building official grants an exception. In this case, you can contact the building official and request an exception to allow the minimum number of water closets to be reduced to one. If the building official approves your request, you can then proceed accordingly.

See Table 1004.1.1 of IBC at the following link:
http://publicecodes.citation.com/icod/ibc/2006f2/icod_ibc_2006f2_10_sec004_par001.ht m?bu=IC-P-2006-000001&bu2=IC-P-2006-000019

See Table 2902.1 of IBC at the following link:
http://publicecodes.citation.com/icod/ibc/2006f2/icod_ibc_2006f2_29_sec002.htm?bu=I C-P-2006-000001&bu2=IC-P-2006-000019

46. Answer: d
Architects often specify Type "X" Gypsum Wallboard for fire resistance. Type "X" Gypsum Board is more expensive than regular gypsum wallboard, and is NOT easier to install than regular gypsum wallboard. It has no recyclability.

47. Answer: a
The ability of water to flow against gravity through concrete floor cracks is called a **capillary action**.

Seepage is the slow escape of a liquid or gas through porous material or small holes.

Saturation is the state or process that occurs when no more of something can be absorbed, combined with, or added.

Leakage is the accidental admission or escape of a fluid or gas through a hole or crack.

48. Answer: a

The image A shows a W shape steel or wide flange.
The image B shows an S shape steel or American standard I beam.
The image C shows a T shape steel or structural tee. IT can be cut from a W shape steel or an S shape steel.
The image D shows a C channel or American standard channel.
The image E shows an equal leg steel angle (there are also unequal leg steel angles).
The image F shows a square tube steel (There are also rectangle or round tube steel shapes or pipes).

There are also combined shapes, such as double angles, a combination section of a C channel and a W shape steel

See the following books for various basic steel shapes:

Steel Construction Manual, Latest edition
American Institute of Steel Construction

OR
Handbook of Steel Construction, Latest edition; and *CAN/CSA-S16-01 and CISC Commentary*
Canadian Institute of Steel Construction

49. Answer: b
    The architect is responsible for this accident because the constructions plans do not show any safety railing around the roof hatch opening.

    ***Note:** Many architects miss the roof hatch safety railing but BOTH OSHA Standards 29 CFR 1910.23 and 29 CFR 1910.27 require it.*

    Make sure you show a standard roof hatch safety railing. It can be a simple note calling out the manufacturer and model number and a simple graphic on your roof hatch detail.

    I know at least three manufacturers who make them. See the links below:
    http://www.simplifiedbuilding.com/keehatch.php
    http://www.freepatentsonline.com/6681528.html
    http://www.4specs.com/s/07/07-7230.html

50. Answer:
    Per *International Mechanical Code* (IMC), if the height of a roof access ladder is over **30** feet, an intermediate landing is required.

    See Section 306.5 of the *International Mechanical Code* (IMC) at the following link:
    http://publiccodes.citation.com/icod/imc/2006f2/icod_imc_2006f2_3_sec006_par007.htm

51. Answer: b and d
    Please note we are looking for statements that are NOT true.

    The following statements are not true and therefore the correct answers:
    - Batt insulation can be installed under the concrete slab. (Batt insulation cannot be installed under the concrete slab. It can be installed under the roof deck, or right above the suspended ceiling.)
    - Rigid insulation is typically attached to the bottom of the roof deck. (Rigid insulation is typically installed above the roof deck.)

    The following statements are true, and therefore the incorrect answers:
    - Batt insulation can attached to the bottom of the roof deck.

• Rigid insulation can be installed under the concrete slab.

52. Answer: d
Building's population has the least impact on the design of elevators.

Accessibility and safety are mandatory requirements for the design of elevators

Number of passengers at peak hour will help the designer to determine the number and size of the elevators needed.

Building's population has some influences on the design of elevators, but is has the least impact on the design of elevators.

53. Answer:
The <u>nominal</u> size of a standard brick in the US is Length (L) × Depth (D) × Height (H) = <u>8" × 4" × 2 5/8" (203 × 102 × 67)</u>, and the <u>nominal</u> size of a standard concrete masonry unit (CMU) in the US is H × D x L = <u>8" × 8" × 16" (203 × 203 × 406)</u>. The actual size is usually about <u>3/8" (10)</u> smaller to allow for mortar joints.

This means:
The <u>actual</u> size of a standard brick in the US is <u>7 5/8" × 3 5/8" × 2 1/4" (193 × 92 × 57)</u>, and the <u>actual</u> size of a standard CMU in the US is <u>7 5/8" × 7 5/8" × 15 5/8" (193 × 193 × 396)</u>.

54. Answer: b and c
Both 19'-4" and 20'-0" fit the CMU block module dimension. The nominal size of a standard concrete masonry unit (CMU) in the US is <u>8" × 8" × 16" (203 × 203 × 406)</u>. Most CMU manufacturers also produce half blocks at the standard size of <u>8" × 8" × 8" (203 × 203 × 203)</u>.

(18'-4") / 8" = 27.5   (This dimensions is not according to CMU block module, and CMU blocks have to be cut.)

(19'-4") / 8" = 29   (This dimensions is according to CMU block module, and CMU blocks do NOT have to be cut.)

(20'-0") / 8" = 30   (This dimensions is according to CMU block module, and CMU blocks do NOT have to be cut.)

(20'-4") /8" = 30.5   (This dimensions is not according to CMU block module, and CMU blocks have to be cut.)

Note: There is a simple rule of thumb to determine if a dimension meets the CMU block module:

*If the dimension has an **odd** number on the feet unit, and you **end up with 4"**, it meets the CMU block module. For example, 1'-4", 3'-4", 5'-4", 7'-4", 9'-4", and 11'-4" all meet the CMU block module.*

*If the dimension has an **even** number on the feet unit, and you **end up with 0" or 8"**, it meets the CMU block module. For example, 2'-0", 2'-8", 4'-0", 4'-8", 6'-0", and 6''-8" all meet the CMU block module.*

55. Answer: b and d
The following are true:
- She needs to use dimensions to locate the work points for the escalator. (An architect needs to refer to escalator manufacturer's brochure, specify the model of the escalator, locate the two work points for the escalator, and make sure there is adequate clearance and headroom below the escalator. The contractor can show the remaining dimensions as part of the shop drawings or submittals.)
- She needs to coordinate with structural and electrical engineers

The following are not true:
- She needs to determine the vertical slope of the escalator. (The vertical slope of an escalator is always set at 30 degree. An architect does NOT determine it)
- She needs to draw and add full dimensions for the escalator so that the contractor can build the escalator accurately. (She needs to use dimensions to locate the work points for the escalator, but she does NOT need to draw and add full dimensions for the escalator. The contractor does the full dimensions as part of the shops drawings or submittals.)

56. Answer: b
The image B shows a Weather Struck brick joint.

The image A shows a Concave brick joint.
The image C shows a Vee brick joint.
The image D shows a Flush brick joint.
The image E shows a Raked brick joint.
The image F shows a Trowel Struck brick joint.

57. Answer: c
Strips made of white polyethylene are often added to the top and side of glass block partition. The purpose of the strips is to provide room for glass block expansion. The strips are called **expansion strips.**

The following are incorrect answers:
- To absorb extra moisture from the glass block partition.
- To protect the glass block partition in an earthquake. (There is some truth to this, but it is NOT the best answer.)
- None of the above

58. Answer: d
The correct answer is "all of the above."

The purpose of sliptrack or slotted deflection track on the top of the full-height metal stud walls is:
- To protect the metal stud walls from wind forces.
- To protect the metal stud walls in an earthquake.
- To provide room for metal stud expansion and contraction.

59. Answer: a
The image A shows a Fink truss.

The image B shows a king post truss.
The image C shows a queen post truss.
The image D shows a scissor truss.
The image E shows a Warren truss.
The image F shows a Howe truss.
The image G shows a Pratt truss.

60. Answer: d
The image D shows a right-hand door.

The image A shows a left-hand reverse door.
The image B shows a right-hand reverse door.
The image C shows a left-hand door.

61. Answer: a, e and f
The following are the federal laws that mandate accessibility to certain historic structures:
- Americans with Disabilities Act
- Architectural Barriers Act
- Section 504 of the Rehabilitation Act

Fair Housing Act is the basic federal laws involve housing accessibility issues:

The following are the basic California laws involve accessibility issues:
- Fair Employment & Housing Act
- Unruh Civil Rights Act

See page 14 of the PDF file for *The Secretary of the Interior's Standards for the Treatment of Historic Properties with Guidelines for Preserving, Rehabilitating Restoring & Reconstructing Historic Buildings* at the following links: http://www.ironwarrior.org/ARE/Historic_Preservation/

62. Answer: d
Providing equivalent exhibits on the first floor for handicap people is the best solution to comply with accessibility requirements in a 2-story historic museum. This solution will NOT alter the historic museum.

Providing a refuge area next to the stair on the second floor will only provide refuge area for emergency, and does not provide accessibility for handicap people.

All the other solutions listed below will alter the historic museum, and therefore are not the correct answers:
- Install an elevator to provide access to the second floor
- Install a lift to provide access to the second floor

63. Answer: a
Not having documentation on the new work is not recommended when working on the entrance of a historic building.

The following are all acceptable:
Stabilizing deteriorated or damaged entrances and porches as a preliminary measure.

Repairing entrances and porches by reinforcing the historic materials using recognized preservation methods.

Replacement: Replacing in kind extensively deteriorated or missing parts.

See pages 38 and 39 of the PDF file for *The Secretary of the Interior's Standards for the Treatment of Historic Properties with Guidelines for Preserving, Rehabilitating Restoring & Reconstructing Historic Buildings* at the following links:
http://www.ironwarrior.org/ARE/Historic_Preservation/

64. Answer: b and c
If some of the exposed beams of an historic building have deflection, the following are recommended:
- Augmenting or upgrading
- Protecting and maintaining the beams and ensuring that structural members are free from insect infestation.

The following are not recommended:
- Leaving them in place without doing anything (This creates a life safety issue.)
- Replacement of <u>all</u> the beams (Repairing or limited replacement of <u>deflected</u> beams is appropriate.)

See pages 42 and 43 of the PDF file for *The Secretary of the Interior's Standards for the Treatment of Historic Properties with Guidelines for Preserving, Rehabilitating Restoring & Reconstructing Historic Buildings* at the following links:
http://www.ironwarrior.org/ARE/Historic_Preservation/

65. Answer: c and d
    The following are not recommended when working on preservation of the interior of a historic building:
    - Using propane to remove paint (Using destructive methods such as propane or butane torches or sandblasting to remove paint or other coatings is not recommended.)
    - Sandblasting of character-defining features

    The following are recommended when working on the interior of a historic building:
    - Installing protective coverings for wall coverings in the corridors
    - Replacement of broken window glazing (Changing the texture and patina of character-defining features through sandblasting or use of abrasive methods to remove paint, discoloration or plaster is not recommended.)

    See pages 45 and 46 of the PDF file for *The Secretary of the Interior's Standards for the Treatment of Historic Properties with Guidelines for Preserving, Rehabilitating Restoring & Reconstructing Historic Buildings* at the following links:
    http://www.ironwarrior.org/ARE/Historic_Preservation/

66. Answer: c
    A new mechanical system is required to make a historic building functional. The following is recommended:
    - Place the mechanical unit on the ground and enclose it with hedges.

    The following are NOT recommended:
    - Place the mechanical unit in the attic and remove a substantial amount of building materials.
    - Place the mechanical unit in an existing masonry unit enclosure and cut through the existing masonry walls.
    - Place the mechanical unit on the roof top and add additional beams and columns to support the unit

    See page 50 of the PDF file for *The Secretary of the Interior's Standards for the Treatment of Historic Properties with Guidelines for Preserving, Rehabilitating Restoring & Reconstructing Historic Buildings* at the following links:
    http://www.ironwarrior.org/ARE/Historic_Preservation/

67. Answer: a
    The following procedure is recommended for historic building:
    - Identifying, retaining, and preserving, stabilizing, protecting and maintaining, repairing, and replacement

    See the PDF file for *The Secretary of the Interior's Standards for the Treatment of Historic Properties with Guidelines for Preserving, Rehabilitating Restoring & Reconstructing Historic Buildings* at the following links:
    http://www.ironwarrior.org/ARE/Historic_Preservation/

68. Answer: b and c

After code research, an architect discovers an exposed wood stair in a historic building has to be fire rated. The following procedure are recommended:
- Upgrading historic stairway so that it is not damaged or obscured
- Adding a new stairway or elevator to meet health and safety codes in a manner that preserves adjacent character-defining features and spaces

The following procedures are NOT recommended:
- Enclosing the wood stair with fire-resistant sheathing required by codes (Covering character-defining wood features with fire-resistant sheathing which results in altering their visual appearance is NOT recommended.)
- Installing sensitively designed fire suppression systems, such as sprinkler systems that result in retention of historic features and finishes (This will NOT meet the exiting codes requirements.)

See page 59 of the PDF file for *The Secretary of the Interior's Standards for the Treatment of Historic Properties with Guidelines for Preserving, Rehabilitating Restoring & Reconstructing Historic Buildings* at the following links: http://www.ironwarrior.org/ARE/Historic_Preservation/

69. Answer: c

The lead-based paint for a historic building starts to peel, chip, craze, or otherwise comes loose. The following procedure is recommended:
- Remove the lead-paint throughout the building and apply a compatible primer and finish paint (Special license, training and/or protection gear and clothing is often required for removing the hazardous lead-based paint)

The following are NOT recommended:
- Leave the paint in place and do nothing (The lead-based paint for a historic building start to peel, chip, craze, or otherwise comes loose. This is a health and safety concern, and can NOT be left in place. If the existing lead-based paint is in good condition, it can be left in place)
- Use the same kind of lead-based paint to repaint the damaged area (The lead-based paint has been banned in construction since late 1970s.)

70. Answer: a and b

According to *The Secretary of the Interior's Standards for the Treatment of Historic Properties*, the following treatments are appropriate for a building individually listed in the National Register:
- Preservation
- Restoration

The following treatments are inappropriate for a building individually listed in the National Register:

- Rehabilitation (It is for buildings that contribute to the significance of a historic district but are NOT individually listed in the National Register)
- Reconstruction (It is for re-creating a vanished or non-surviving building with new materials, primarily for interpretive purposes.)

See pages 1 and 2 of the PDF file for *The Secretary of the Interior's Standards for the Treatment of Historic Properties with Guidelines for Preserving, Rehabilitating Restoring & Reconstructing Historic Buildings* at the following links:
http://www.ironwarrior.org/ARE/Historic_Preservation/
http://www.nps.gov/hps/tps/standguide/overview/choose_treat.htm

71. Answer: a
Cast-in-place concrete beam-and-slab system is the most appropriate structural system for a lab building that is sensitive to vibration.

The following systems are not as good as cast-in-place concrete beam-and-slab system:
- Heavy timber construction with panelized floor
- Lightweight concrete over metal deck over steel joists
- 4" gypsum concrete topping slab over wood deck over wood joists

72. Answer: d
The following is a proper action for the architect:
- Denies the change order.

The following are improper action for the architect:
- Approves the change order and back charges the cost to the structural engineer.
- Approves the change order and submits a claim to her professional liabilities insurance company and to the structural engineer's professional liabilities insurance company.
- Negotiates with the contractor to reduce the amount of change order, obtains the owner's approval and then approves the change order.

Per Article 4.2.7 of A201–2007, General Conditions of the Contract for Construction, an architect review submittals including shop drawings for conformance with information given and the design concept expressed in the construction documents. The review is NOT for the purpose of determining the accuracy and completeness of other details such as dimensions and quantities.

You can find a FREE PDF file of commentary for AIA document A201–2007, General Conditions of the Contract for Construction, at the following link:
http://www.aia.org/contractdocs/aiab081438

73. Answer: d
The principal AIA documents A201 family includes:
- Some documents in the A series, like agreement between owner and contractor and agreement between contractor and subcontractor, some documents in the B series,

like agreement between owner and architect, and some documents in the C series, like agreement between architect and consultant

See Introduction of A201–2007, General Conditions of the Contract for Construction,

You can find a FREE PDF file of commentary for AIA document A201–2007, General Conditions of the Contract for Construction, at the following link: http://www.aia.org/contractdocs/aiab081438

74. Answer: a and c
The following are appropriate for extinguishing a fire in a cell phone equipment room:
- Dry ice
- Dry chemicals

The following and any other substance containing water are inappropriate for extinguishing a fire in a cell phone equipment room:
- Water
- A combination of water, carbon dioxide and dry chemicals

75. Answer: c
Image C shows the correct location of perforated pipe. The main purpose of the perforated pipe is to collect the water from the retaining portion of soil behind the retaining wall, and slopes and drains the water to an outlet away from wall.

76. Answer: b
The previous image shows a rigid frame: the connections between the beam and the columns are rigid.

The following are incorrect answers:
- Column and beam (The connections between the beam and the columns are flexible, and the column-and-beam assembly is NOT able to resist lateral forces.)
- Space frame (a planar unit consists of rigid, short linear members assembled into a three-dimensional triangle pattern)
- Prefabricated column and beam (The connections between the beam and the columns are flexible, and the column-and-beam assembly is NOT able to resist lateral forces.)

77. Answer: a
Waterstop is the correct term for the preformed synthetic rubber labeled as letter A in previous image.

The following are incorrect answers:
- Moisture barrier (It is a water-resistant membrane placed under the slab or in the warm side of the walls).
- Control joint (It is a pre-determined concrete joint for concrete to crack along it.)
- Isolation joint (It is a joint that completely separate the concrete or the building.)

78. Answer: c
    The image C shows the correct placement of water resistant membrane.

    The images A, B and D show the incorrect placement of water resistant membrane:
    The water resistant membrane is not placed deep enough, and water and moisture can still penetrate through the retaining wall and gets into the building.

79. Answer: c
    The image C shows a Sash door.

    The image A shows a French door.
    The image B shows a Glass door.
    The image D shows a Panel door.

80. Answer: c
    Firestopping is the correct term for A in the previous image. The **fireproofing** around the wide flange steel beam is a major clue for the correct answer.

    Firestopping is required to seal off the gap between the wall and the edge of each floor. It has to be secured to the structure. Firestopping can consist of metal lath and plaster, steel plate and grout, or mineral wool safing.

81. Answer: a
    In a fast-track, one-story project, roof framing system design is most likely a critical path item because it involves coordination with HVAC system and takes the longest time and will affect the progress of other elements.

    The following will all take less time:
    • HVAC system design
    • Interior finish selection
    • Door selection

82. Answer: a and b
    Grade beams are underground beams connecting isolated footings

    The following situation shall utilize grade beams:
    • At site with expansive soils (Grade beams can transfer the building load more evenly and prevent uneven settlement)
    • At storefront with no solid walls (This situation typical requires a moment frame system, including grade beams because storefront with no solid walls does NOT have adequate shear walls to resist the lateral force.)

    The following situation shall NOT utilize grade beams:
    • At site with underground rock

    The following situation may NOT utilize grade beams:

- At site in an inland area

83. Answer: b
Quirk miter is the correct term for the joint shown in image A.

The following are some basic veneer stone corner joints:
- Corner "L" (Image B)
- Slip corner (Image C)
- Butt joint (Image D)

84. Answer: a
Cripple studs is the correct term for A in the image.

The following are incorrect answers:
- King posts (A term for trusses, and NOT for studs at all)
- Transom studs (Invented distractor)
- Top studs (Invented distractor)

85. Answer: c
Square edge tongue and groove is the correct term for the wood siding A in the image.

The following are incorrect answers:
- Bevel (Wood siding B)
- Shiplap (Wood siding C)
- Channel rustic (Wood siding D)

B. Mock Exam Solution: BDCS Accessibility/Ramp Vignette

1. Major criteria, overall strategy, and tips for accessibility/ramp vignette

1) Set the minimum dimension for the **landing** and **width of the ramp** to be **5'-0"** (1524) since the ramp for this vignette typically HAVE to change direction. Set the **width of the stair** to 5'-0" (1524) to match that of the ramp.
2) Make sure you show the **proper elevations** for the landings.
3) Do NOT forget to extend the ramp non-continuous handrails horizontally at least **12 inches** (305) beyond the **top and bottom** of the ramp run.
4) Do NOT forget to extend the stair non-continuous handrails horizontally at least **12 inches** (305) beyond the **top and bottom** risers.
5) Exit doors shall swing in the direction of egress travel.
6) Maintain enough clearance for both the push side and the pull side of the egress door.

2. Step-by-step solution for the mock exam accessibility/ramp vignette using the official NCARB BDCS practice program

1) Open the NCARB graphic program, and find a large blank area.

2) Since we have all the critical dimensions for the buildings, we can use the **Sketch** tool of NCARB graphic program to draw a background outline to define the major spaces that we will work on (Figure 4.1).

3) Setting the **length and width of the ramp as well as** the **minimum landing dimension:** Since the elevation difference between the lobby and the upper level is 28" (711), at 1:12 slope, the total length of the ramp is 28' (8534).

   Since the lobby size is 22'-0" x 28'-0" (6706 x 8534), and you HAVE to **change the ramp direction** for this vignette since the total length of the ramp is 28' (8534) and you need room for the landings. You have to set the ramp width and the minimum landing dimension to **5'-0"** (1524) since the ramp changes direction.

4) Click on Cursor to change the cursor to full-screen cursor; Use **Draw > Landing** to draw a **landing #1** covering the entire width of the lobby, and landing width is 5'-0" (1524, Figure 4.2). You need to click three times: the **first click** sets the upper left hand corner of the landing at the upper left hand corner of the lobby; **second click** sets the upper right hand corner of the landing at the upper right hand corner of the lobby; **third click** sets the vertical dimension of the landing to 5'-0" (1524). On the lower left hand corner of the screen, you can see the dimensions of the rectangle is x: 22'-0", y: 5'-0" (x: 6706, y: 1524).

5) Click on **Set Elevation** to bring up the elevation dialogue box, use the up arrow to set the **landing #1** elevation to 28" to match the elevation of the upper exit corridor (Figure 4.3).

   Note: There is a defect in the NCARB graphic program. You cannot use the left-side up arrow to directly set the elevation to 28". You have to set it to 29" first, and then use the right-side down arrow to reduce the elevation to 28.95", 28.85"… all the way to 28".

6) Use **Draw > Landing** to draw a 10'-0" x 5'-0" (3048 x 1524) **landing #2** at the lower left-hand corner of the lobby. **Set Elevation** to 10" (Figure 4.4).

7) Use **Draw > Ramp** to draw an 18'-0" (5486) long x 5'-0" (1524) wide ramp #1 connecting **landing #1** and landing #2. Pay attention to the **direction of travel** for the ramp. Since the ramp #1 is 18'-0" (5486) long, the elevation of the landing #2 should be 18" (457) lower than landing #1 = 28"- 18" = 10" (711-457=254).

8) Click on **Set Elevation** to bring up the elevation dialogue box, use the up arrow to set the **landing #2** elevation to 10" (254, Figure 4.5).

9) Use **Draw > Ramp** to draw a roughly 10'-0" (3048) long x 5'-0" (1524) wide ramp #2 connecting **landing #2** and **the lobby floor**. Pay attention to the **direction of travel** for

the ramp. Use **Zoom** to zoom in, and use Move Adjust to adjust ramp #2 to the accurate 10'-0" (3048) dimension if necessary (Figure 4.6).

10) Since maximum riser height shall be 7 inches (178) and minimum riser height shall be 4 inches (102), we are going to use 7" high riser, and we need 28"/7" = 4 risers, or 3 treads. We use 11" (279) deep treads, so the total stair length = 3 x 11" = 33" or 2'-9" (838). We set the stair width to 5'-0" (1524).

Use **Draw > Stair > Direction (of Stair) > # of Risers > 4** to draw the stair. Pay attention to the **direction of travel** for the stair (Figure 4.7).

11) Use **Draw > Railing** to draw all the railings (Figure 4.8).

Do NOT forget to extend the ramps' non-continuous handrails horizontally at least **12 inches** (305) beyond the **top and bottom** of the ramp run.

Do NOT forget to extend the stair's non-continuous handrails horizontally at least **12 inches** (305) beyond the **top and bottom** risers.

The railings at the corner of the upper exit corridor have to turn 90 degrees and extend the **12 inches**(305)**.**

12) You can use **Sketch > Rectangle** to draw some temporary rectangles to assist you locate the end of the 12" (305) horizontal extension.

Use **Zoom** to zoom in and **Move, Adjust** to fine tune the railings.

After you finish drawing the railing and related extensions, you can use **Erase** to delete the temporary rectangles to make the lobby area read better (Figure 4.9).

*Note: You need to click on **Erase**, click on the items you want to delete, and then click on **Erase** again to erase them.*

13) Use **Sketch > Rectangle** to draw an 8'-0" x 5'-0" (2438 x 1524) rectangle at the end of the upper exit corridor to locate the full-height wall.

14) Use **Draw > Full Height Wall** to draw a full height wall at the upper exit corridor.

15) We are going to add a pair of 3'-0" (914) doors to the new wall we just draw. You can use **Sketch** to draw two temporary rectangles to locate the doors. Use **Draw > Door > (Select proper door swing direction) > 36"** (914) to draw a door at the full height wall at the upper exit corridor. Repeat the process and draw another door to form a pair door. Make sure the door swing direction matches the direction of egress (Figure 4.10). This completes the solution.

16) To create the final solution read better, I use a **FREE** program called Scribus to overlap Figure 3.20, Figure 4.9, and Figure 4.10 to create a new figure (Figure 4.11).

If you are interested in learning how to use **Scribus**, you can read my other book:

Using FREE Scribus Software to Create Professional Presentations: *Book Covers, Magazine Covers, Graphic Designs, Posters, Newsletters, Renderings, and More (Full Color Edition)*

ISBN: 9780984374151

It is available at http://www.GreenExamEducation.com

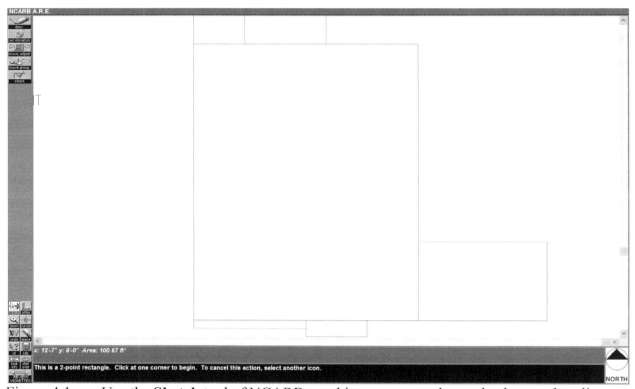

Figure 4.1   Use the **Sketch** tool of NCARB graphic program to draw a background outline to define the major spaces that we will work on.

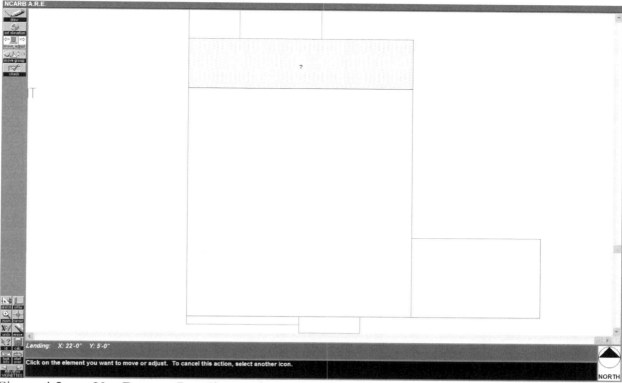

Figure 4.2   Use **Draw** > **Landing** to draw a **landing #1** covering the entire width of the lobby and landing width is 5'-0" (1524).

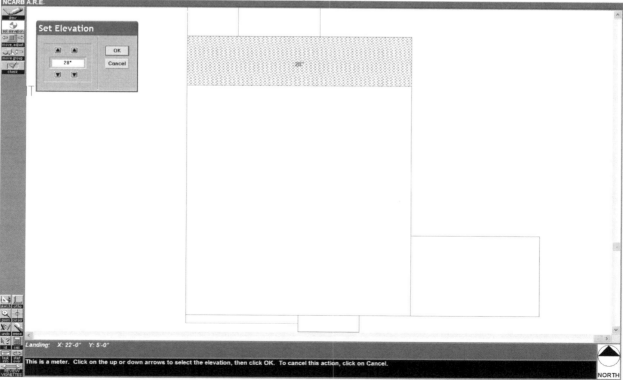

Figure 4.3   Click on **Set Elevation** to bring up the elevation dialogue box, use the up arrow to set the **landing #1** elevation to 28" (711).

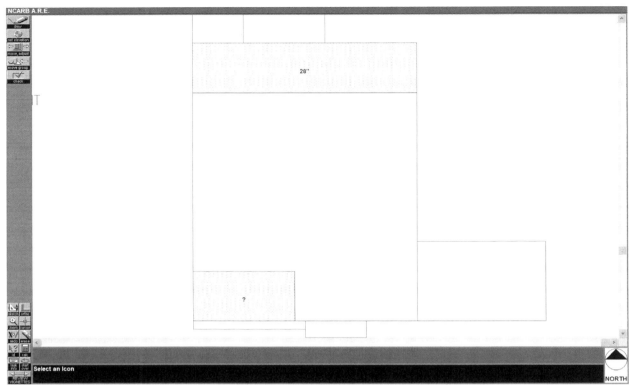

Figure 4.4   Use **Draw** > **Landing** to draw a 10'-0" x 5'-0" (3024 x 1524) **landing #2** at the lower left-hand corner of the lobby. **Set Elevation** to 10" (254).

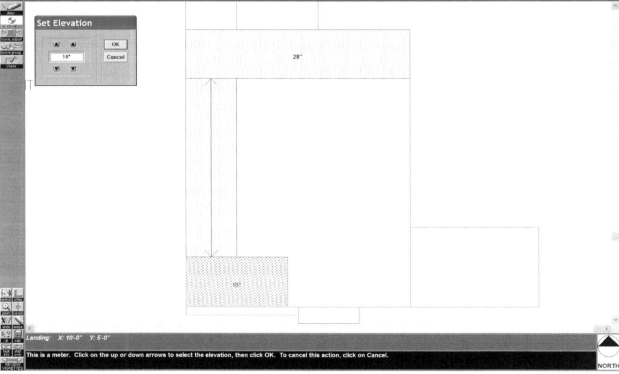

Figure 4.5   Use **Draw** > **Ramp** to draw a 18'-0" (5486) long x 5'-0" (1524) wide ramp #1 connecting **landing #1** and **landing #2**.

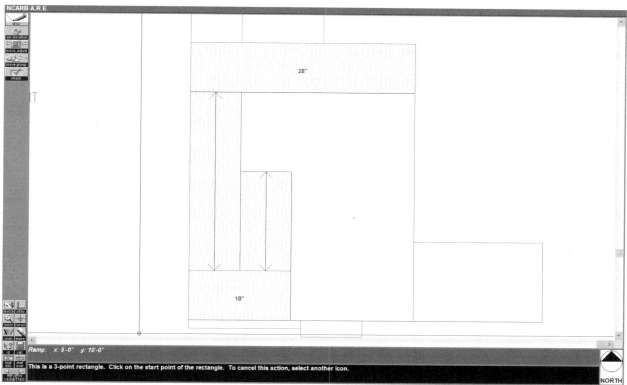

Figure 4.6   Use **Draw > Ramp** to draw a roughly 10'-0" (3024) long x 5'-0" (1524) wide ramp #2 connecting **landing #2** and **the lobby floor**.

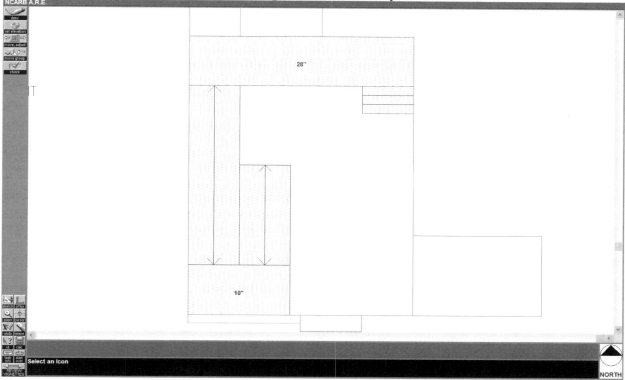

Figure 4.7   Use **Draw > Stair > Direction (of Stair) > # of Risers > 4** to draw the stair. Pay attention to the **direction of travel** for the stair.

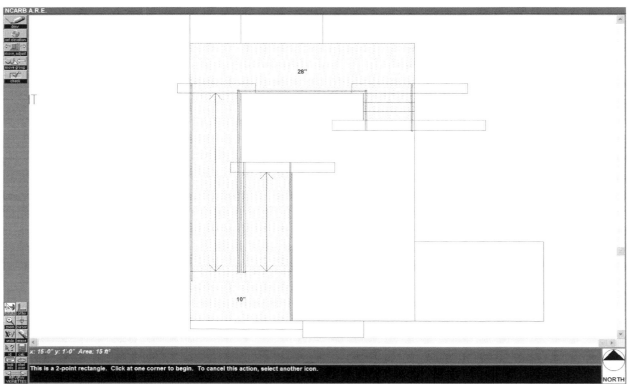

Figure 4.8   Use **Draw** > **Railing** to draw all the railings.

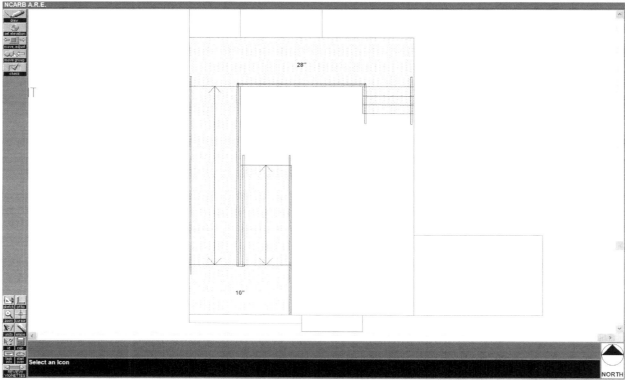

Figure 4.9   Use **Erase** to delete the temporary rectangles to make the lobby area read better.

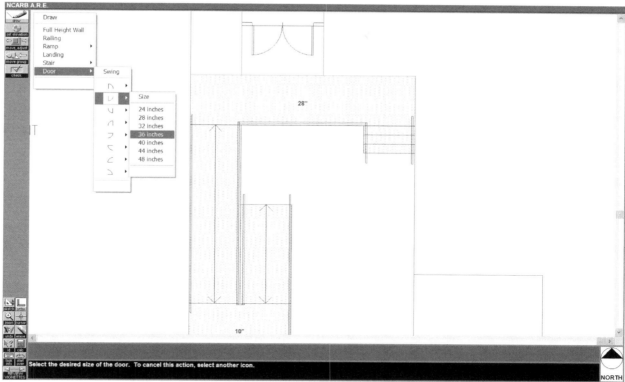

Figure 4.10   Use **Draw > Door > (Select proper door swing direction) > 36" (914)** to draw a door at the full height wall at the upper exit corridor.

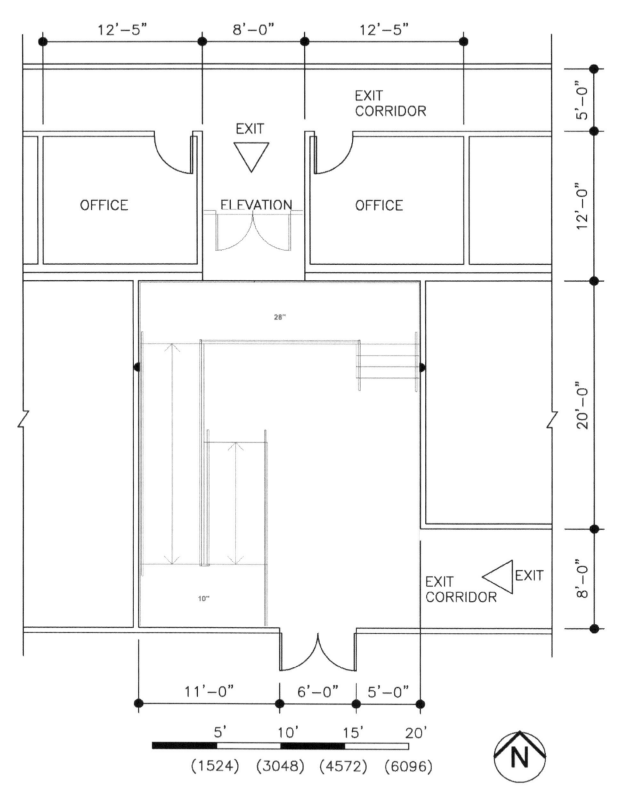

Figure 4.11   Final solution: Overlap Figure 3.20, Figure 4.9, and Figure 4.10 to create a new
                figure.

3. Notes on mock exam vignette traps

Several **common errors** or **traps** into which you may fall:

1) The **width of the ramp** is less than **5'-0"** (1524).
2) The minimum dimension for the **landing** is less than **5'-0"** (1524).
3) Missing the **elevations** for the landings.
4) Forgetting to extend the ramp non-continuous handrails horizontally at least **12 inches** (305) beyond the **top and bottom** of the ramp run.
5) Forgetting to extend the stair non-continuous handrails horizontally at least **12 inches** (305) beyond the **top and bottom** risers.
6) Forgetting to **add railings for landing #1**.
7) Forgetting to draw the **full-height wall** and the related doors.
8) Exit door swings in the wrong direction.
9) Not enough clearance for the pull side of the egress door.

4. A summary of the critical dimensions

1) The **minimum** dimension for the **landing**: 5'-0" (1524)
2) **Ramp width:** 5'-0" (1524)
3) **Stair width:** 5'-0" (1524)
4) **Treads**: 3 @ 11" = 33" (838)
5) **Risers**: 4 @ 7" = 28" (711)
6) Handrail **extension** at **top and bottom** of the ramp runs/Stair: 12" (305)
7) **Clearance** for the **pull side** of the egress door: 24" and 5'-0" x 5'-0" (610 and 1524 x 1524)
8) **Door Size**: A pair of 3'-0" (914) doors

C. Mock Exam Solution: BDCS Stair Design Vignette

1. Major criteria, overall strategy and tips for stair design vignette

1) Show proper elevations of all landings.
2) Show proper elevations of top and bottom of all stair runs.
3) Do NOT forget to extend the ramp non-continuous handrails horizontally at least **12 inches** (305) beyond the **top and bottom** risers.
4) Make sure there is adequate headroom (80" or 2032).
5) The stair should NOT block the egress door and the lobby door on the ground floor.
6) Minimum stair width 44" (1118), subject to occupant load width increase.
7) Show the 30" x 48" (762 x 1219) area of refuge, and the 24" and 5'-0" x 5'-0" (610 and 1524 x 1524) clearance space at the pull side of the egress door. Stair width increased to **48"** (1219) **clear between handrails** when area of refuge is required.
8) The width of a landing shall NOT be less than the width of the stair.
9) Thickness of structure under the stair is 12" (305).
10) Start the stair design from the second floor.

11) By reviewing the ground floor plan and second floor plan, we decide to run the major stair runs along the top wall and toward the left toward, and then go through a landing #2, turn right, continue to a landing right outside the storage room, and then turn left again to the first floor. We need to make sure to provide adequate headroom under this landing because people need to go under to access the exit door. This layout will also make it possible for the storage room to use the same stair for access, and is probably the ONLY solution.

2. Step-by-step solution for the mock exam stair design vignette using the official NCARB BDCS practice program

1) Open the NCARB graphic program, and find a large blank area.

2) Since we have all the critical dimensions for the floor plans, we can use the **Sketch** tool of NCARB graphic program to draw a background outline to define the major spaces that we will work on for both the first and second floors (Figure 4.12).

3) Setting the **stair width** and **minimum landing width**:
   - **Minimum** stair width: 44" (1118)

   - The exit width determined by the calculation based on **occupant load**:

Building Level	Total Occupants Load	Number of Exits	Exit Width
Ground Floor	350	3	(350/3) x 0.3 = 35" (889)
Janitor	8	1	8 x 0.3 = 2.4" (61)
Second Floor	160	2	(160/2) x 0.3 = 24" (610)

   Per NCARB program, the exit width shall be based on the individual floor with the largest occupant load, so, the exit width determined by the calculation based on occupant load is 35" (889). This means the **occupant load** of each level will not require a stair width larger than the 44" (1118) minimum.

   - Since we do have an **area of refuge** for this vignette, the stair width needs to be at least 56" or 1422 (48" or 1219 CLEAR because of the area of refuge, **plus** an 8" or 203 allowance for handrails on both sides). We set the **stair width** and **minimum landing width** as **5'-0"** (1524) to accommodate the railing, and simplify the risers and treads calculations. Setting the stair width to 5'-0" (1524) gives 2" (51) extra on each side of the stair, and gives us more room to draw the railings.

Note: NCARB throws in the 44" (1118) minimum stair width requirement at the beginning of the program, and place the 48" (1219) CLEAR stair width requirement because of the area of refuge at the end of the program. This is to make sure you can coordinate various criteria, and make sure you are patient enough to read the entire program.

4) We start the stair design from the second floor:
Use **Layer** > **Floor Selection** > **2** to turn on only the second floor layers.

5) Use **Sketch** > **Rectangle** to draw a 5'-0" x 5'-0" (1524 x 1524) rectangle to show the clearance space for the second floor door.

Use **Sketch** > **Rectangle** to draw a 4'-0" x 2'-6" (1219 x 762) rectangle to show the 48" x 30" (1219 x 762) area of refuge near the second floor door (Figure 4.13).

6) Use **Draw** > **Landing** to draw a 5'-0" x 8'-0" (x: 5'-0", y: 8'-0") or 1524 x 2438 landing #1 on the upper-right hand corner of the space to cover both the area of refuge and the 5'-0" x 5'-0" (1524 x 1524) clearance space for the second floor door. Click on **Elevation** to bring up the elevation dialogue box, use the **up arrow** to set the elevation of landing #1 to 13'-5" (4089) to match the elevation of second floor (Figure 4.14).

7) Calculate **how many risers we need for the stair runs**:
Per the program, the elevation at the storage level is 2'-4" or 28" (711), and ground floor elevation is 0'-0". The elevation at the second floor is 13'-5" or 161" (4089).

The riser height has to be 7" (178) maximum, and 4" (102) minimum. Let us use 7" (178) high risers with 11" (279) deep tread:

161"/7 = **23 risers (or 22 treads)** needed between the first floor and the second floor.

28"/7 = **4 risers (or 3 treads)** needed between the first floor and the storage level.
The total length of 3 treads = 3 x 11 = 33" = 2'-9" or 838 (stair #3)

Based on our layout, people have to go under landing #2 to access the exit to the sidewalk. The **bottom** of landing #2 has to meet the headroom clearance of 80" or larger, and the **top** of landing #2 elevation has to be equal or larger than the headroom clearance + the landing structural thickness = 80" + 12" = 92" (2337). We choose the top of landing #2 elevation to be 98" or 8'-2" (2489) so that it can be easily divided by the 7" (178) riser height.

The elevation of the second floor - the elevation of landing #2 = 161" – 98" = 63"
63" /7 = **9 risers (or 8 treads)** needed between the second floor and the landing #2
The total length of 8 treads = 8 x 11 = 88" = 7'-4" or 2235 (stair #1)

23 – 4 – 9 = **10 risers (or 9 treads)** needed between the landing #2 and the storage level
The total length of 9 treads = 9 x 11 = 99" = 8'-3" or 2515 (stair #2)

8) Use **Draw** > **Stairs** to draw a 7'-4" (2235) long and 5'-0" (1524) wide stair #1 with 9 risers to the left of landing #1. Pay attention to the direction of travel for the stair. See

previous calculations for the number of risers and treads (Figure 4.15). Use **Zoom** to zoom in, and **Draw, Adjust** to adjust the landing size to align with the top of stair #1 if necessary.

9) Use **Draw > Landing** to draw a 5'-0" x 18'-6" (x: 5'-0", y: 18'-6") or 1524 x 5639 landing #2 span from the top wall to the bottom wall and next to stair #1 (Figure 4.16).

10) Use **Sketch > Rectangle** to draw a 5'-0" (1524) high temporary rectangle (y: 5'-0") to hold the space needed for the stair #3 between the between the first floor and the storage level. This will help us locate the stair #2 at the next step.

11) Use **Draw > Cut Stair** to draw an 8'-3" (2515) long and 5'-0" (1524) wide **(cut) stair #2** with 10 risers to the right of landing #2. Pay attention to the direction of the cut stair symbol. See previous calculations for the number of risers and treads (Figure 4.17).

   Notes:
   *Several key points for using the **Cut Stair** tool:*
   * *When using the cut stair tool, draw the flight of the cut stairs from landing to landing or from landing to the ground floor.*
   * *The flight of the cut stairs has to show up on both the first and the second floor. You need to draw the same flight twice: once for the first floor and the other for the second floor.*
   * *Make sure ALL the data for the same flight of the cut stairs is the same for both floors: top and bottom elevations, depths of treads, and number of risers.*
   * *Extend the railing 1 to 2 clicks over the break line.*

12) Click on **Elevation** to bring up the elevation dialogue box, use the **up arrow** to set the elevation of top of stair #1 to 13'-5" (4089) to match the elevation of landing #1.

   Set the elevation of landing #2 to 8'-2" (2489). See step 7) for calculation of elevation of landing #2.

   Set the elevation of bottom of stair #1 to 8'-2" (2489) to match the elevation of landing #2.

   Set the elevation of top of stair #2 to 8'-2" (2489) to match the elevation of landing #2.

   Set the elevation of bottom of stair #2 to 2'-4" (711) to match the elevation of the storage room (Figure 4.18).

13) Use **Draw > Railing** to draw all the railings for the second floor plan.

   Do NOT forget to extend the ramp non-continuous handrails horizontally at least **12 inches** (305) beyond the **top and bottom** risers.

   For the cut stair #2, extend the railing 1 to 2 clicks over the break line.

You can use **Sketch** > **Rectangle** to draw some temporary rectangles to assist you to locate the end of the 12" (305)railing extensions (Figure 4.19).

Use **Zoom** to zoom in, and **Move, Adjust** to fine turn the railings if necessary (Figure 4.20).

14) We start the stair design from the first floor:
Use **Layer** > **Floor Selection** > **1** to turn on only the first floor layers.

15) Use **Draw** > **Landing** to draw a 5'-0" x 10'-0" (x: 5'-0", y: 10'-0") or 1524 x 3048 landing #3 on the lower-right hand corner of the space (Figure 4.21).

16) Use **Draw** > **Cut Stair** to draw an 8'-3" (2515) long and 5'-0" (1524) wide **(cut) stair #2** with 10 risers to the left of landing #3. Pay attention to the direction of the cut stair symbol. See previous calculations for the number of risers and treads (Figure 4.22).

*Note: We have a conflict here: Stair #2 on the **first** floor layer is overlapping with landing #2 for a tread depth or 11" (279). Stair #2 on the **second** floor layer is also overlapping with landing #3 for a tread depth or 11" (279). In order to resolve this conflict, we'll go back to level 2 and move all the elements 11" (279) to the left, and then increase the width of landing #1 by 11" (279).*

17) Use **Layer** > **Floor Selection** > **2** to turn on only the **second** floor layers.

18) Click on **Move Group** select stair #1, cut stair #2 (at the 2nd level layer), landing #2 and all the related railings, click on **Move Group** again, and move all the highlighted items 11" (279) to the left. You can align the right edge of landing #2 with the left edge of cut stair #2 below (at the 1st level layer) (Figure 4.23).

19) Use **Move, Adjust** to increase the width of landing #1 to 5'-11" (x: 5'-11") or 1803, adjust the railings accordingly (Figure 4.24).

*Note: We could have drawn the landing #1 as 5'-11" x 8'-0" (x: 5'-11", y: 8'-0") or 1803 x 2438 instead of 5'-0" x 8'-0" (x: 5'-0", y: 8'-0") or 1524 x 2438 from the very beginning, and avoided steps 18) and 19) above, but you would not have know the reasons. Steps 18) and 19) are very good exercises for the **Move Group** tool for you to practice in case you need to do a major revision like this in real exam.*

*Of course, in the real exam, you can set the landing #1 as 5'-11" x 8'-0" (x: 5'-11", y: 8'-0") instead of 5'-0" x 8'-0" (x: 5'-0", y: 8'-0") **from the very beginning** to save time.*

20) Use **Layer** > **Floor Selection** > **1** to turn on only the first floor layers.

21) Use **Draw** > **Stairs** to draw a 2'-9" (838) long stair #3 with 4 risers (3 treads) to the left of landing #3. Pay attention to the direction of travel for the stair (Figure 4.25).

22) Click on **Elevation** to bring up the elevation dialogue box, use the **up arrow** to set the elevation of landing #3 to 2'-4" (711) to match the elevation of janitor room.

Set the elevation of top of stair #2 to 8'-2" (2489) to match the elevation of landing #2.

Set the elevation of bottom of stair #2 to 2'-4" (711) to match the elevation of landing #3 and the storage room.

Set the elevation of top of stair #3 to 2'-4" (711) to match the elevation of landing #3.

Set the elevation of bottom of stair #3 to 0'-0" to match the elevation of ground floor (Figure 4.26).

23) Use **Draw > Railing** to draw all the railings for the first floor plan.

Do NOT forget to extend the ramp non-continuous handrails horizontally at least **12 inches** beyond the **top and bottom** risers.

You can use **Sketch > Rectangle** to draw some temporary rectangles to assist you to locate the end of the 12" (305) railing extensions.

For the cut stair #2, extend the railing 1 to 2 clicks over the break line (Figure 4.27).

Use **Zoom** to zoom in, and **Move, Adjust** to fine turn the railings if necessary.

24) After you finish all railings, Use **Zoom** to zoom out, and delete the temporary rectangles. This is you final solution (Figure 4.28 and Figure 4.29).

Figure 4.12   Use the **Sketch** tool of NCARB graphic program to draw a background outline to define the major spaces that we will work on.

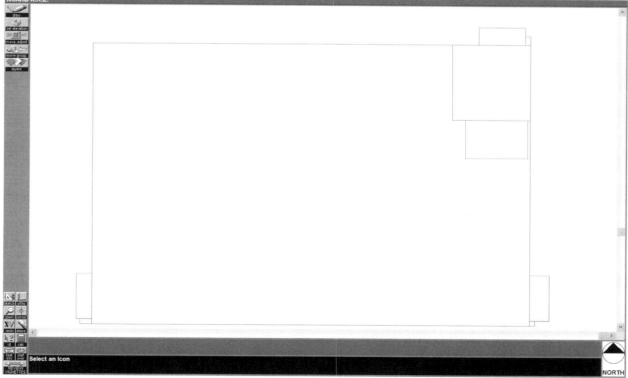

Figure 4.13   Use **Sketch > Rectangle** to draw a 4'-0" x 2'-6" (1219 x 762) rectangle to show the 48" x 30" (1219 x 762) area of refuge near the second floor door.

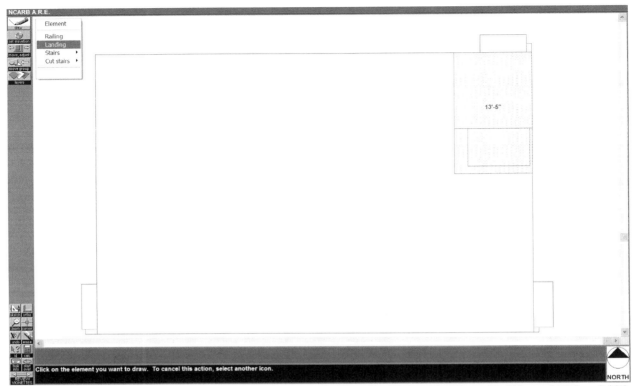

Figure 4.14   Use **Draw** > **Landing** to draw a 5'-0" x 8'-0" (x: 5'-0", y: 8'-0") or 1524 x 2438 landing #1 on the upper-right hand corner of the space.

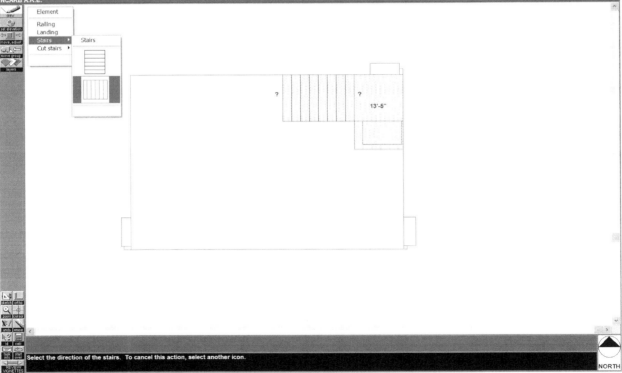

Figure 4.15   Use **Draw** > **Stairs** to draw a 7'-4" (2235) long stair #1 with 9 risers to the left of landing #1.

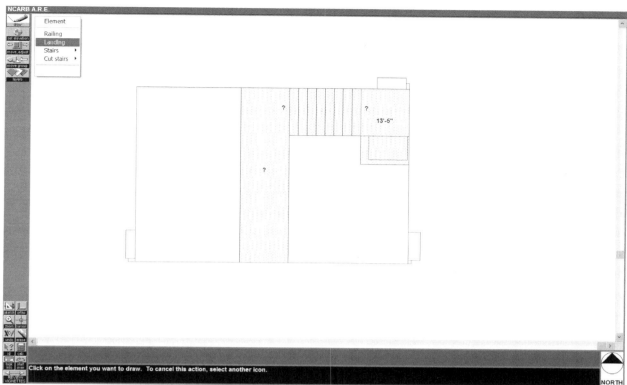

Figure 4.16   Use **Draw > Landing** to draw a 5'-0" x 18'-6" (x: 5'-0", y: 18'-6") or 1524 x 5639 landing #2 span from the top wall to the bottom wall and next to stair #1.

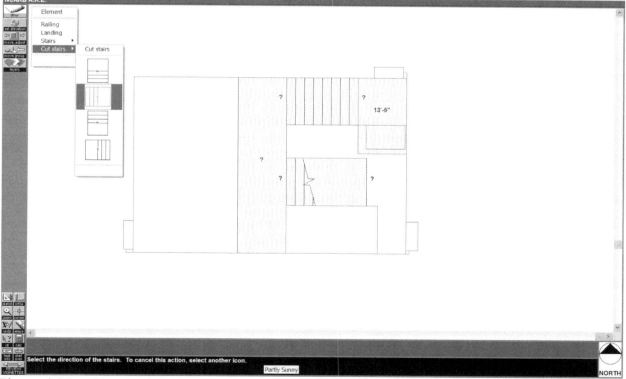

Figure 4.17   Use **Draw > Cut Stair** to draw an 8'-3" (2515) long and 5'-0" (1524) wide **(cut) stair #2** with 10 risers to the right of landing #2.

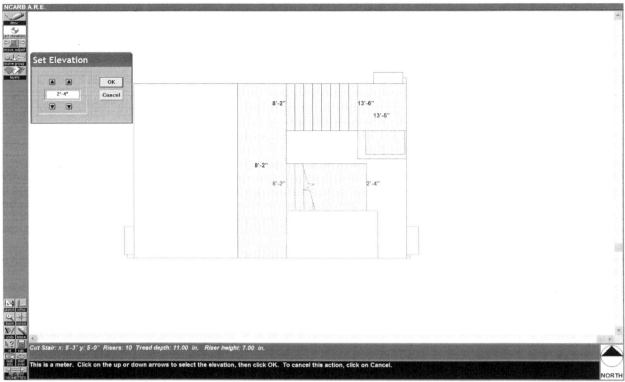

Figure 4.18   Click on **Elevation** to bring up the elevation dialogue box to set the elevations of the landings and the top and bottom of stairs for the second floor plan.

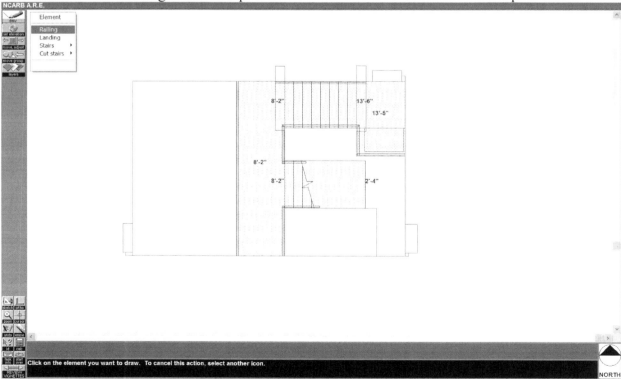

Figure 4.19   Use **Draw > Railing** to draw all the railings for the second floor plan.

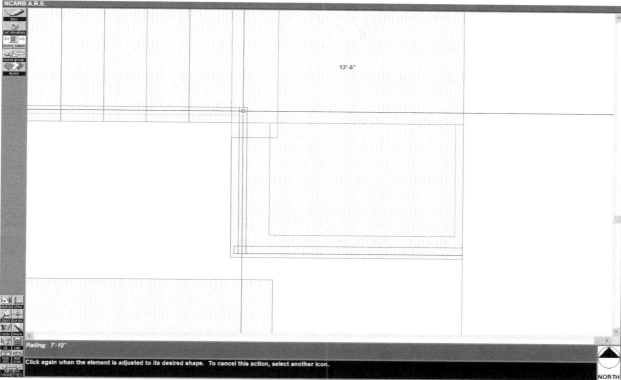

Figure 4.20   Use **Zoom** to zoom in, and **Move, Adjust** to fine turn the railings if necessary.

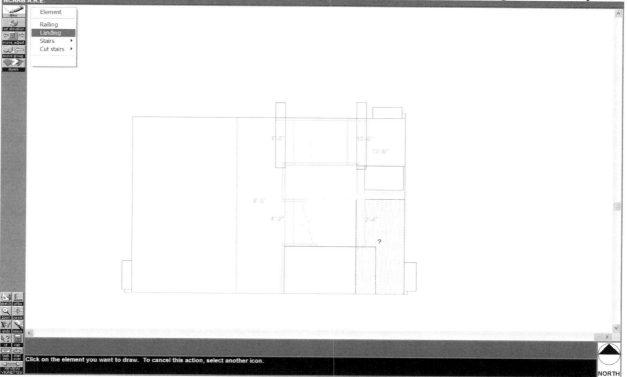

Figure 4.21   Use **Draw > Landing** to draw a 5'-0" x 10'-0" (x: 5'-0", y: 10'-0") or 1524 x 3048 landing #3 on the lower-right hand corner of the space on the **first** floor layer.

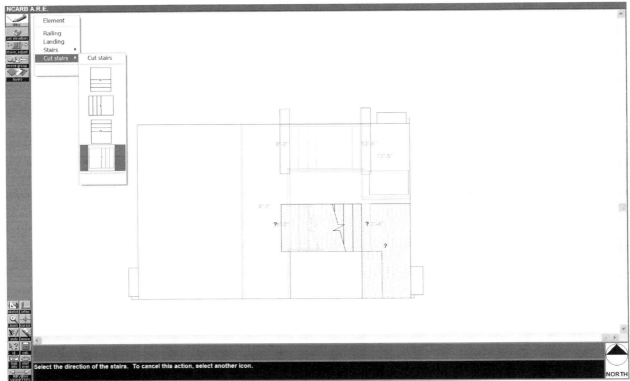

Figure 4.22   Use **Draw > Cut Stair** to draw an 8'-3" (2515) long and 5'-0" (1524) wide **(cut) stair #2**.

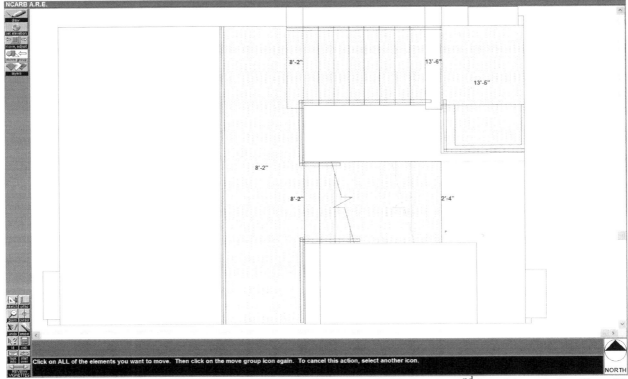

Figure 4.23   Use **Move Group** to move stair #1, cut stair #2 (at the 2nd level layer), landing #2 and all the related railings 11" to the left.

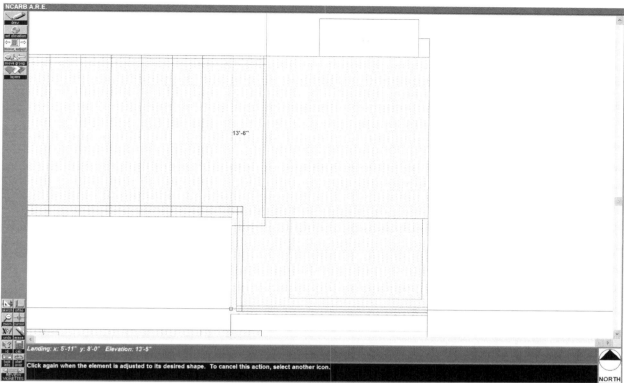

Figure 4.24   Use **Move, Adjust** to increase the width of landing #1 to 5'-11" (x: 5'-11") or 1803, adjust the railings accordingly.

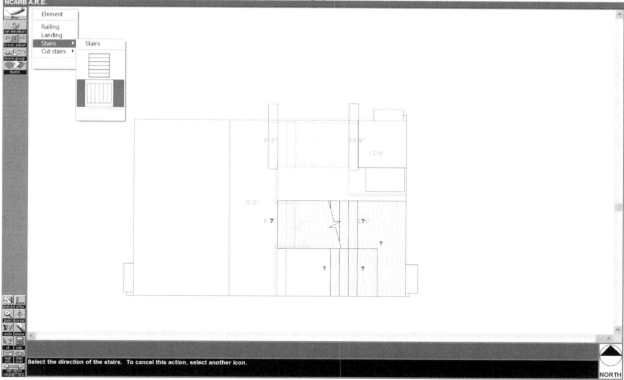

Figure 4.25   Use **Draw > Stairs** to draw a 2'-9" (838) long stair #3 with 4 risers (3 treads) to the left of landing #3. Pay attention to the direction of travel for the stair.

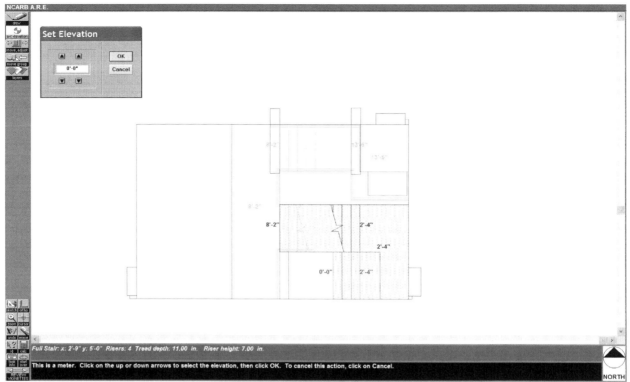

Figure 4.26   Click on **Elevation** to bring up the elevation dialogue box to set the elevations of the landings and the top and bottom of stairs for the first floor plan.

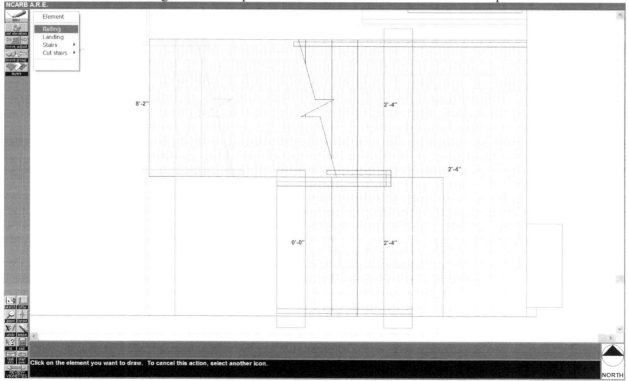

Figure 4.27   Use **Draw > Railing** to draw all the railings for the first floor plan.

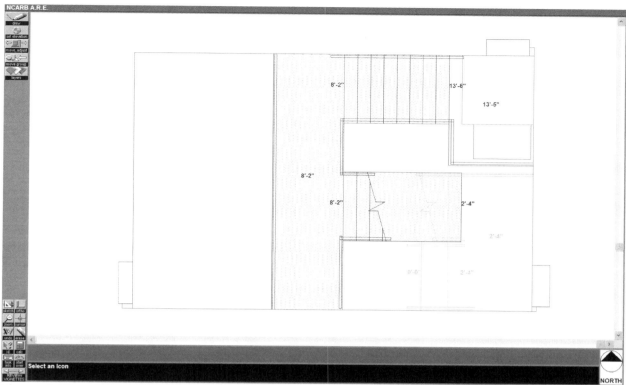

Figure 4.28   The final solution for the second floor plan.

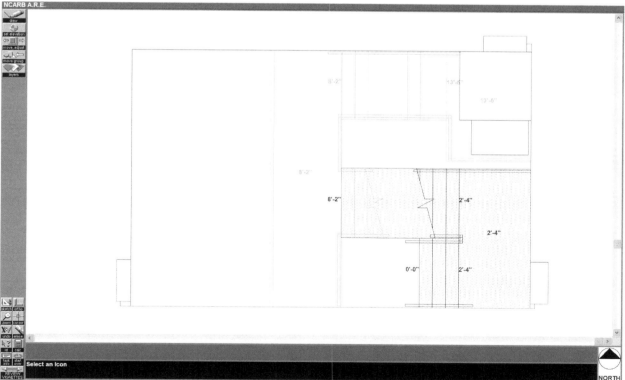

Figure 4.29   The final solution for the first floor plan.

3. Notes on mock exam vignette traps

Several **common errors** or **traps** into which you may fall:
1) Forgetting to show **proper** elevations of all landings.
2) Elevations of landings **do not match** the adjacent top or bottom of the stair runs, or the elevations of the floors.
3) Forgetting to extend the ramp non-continuous handrails horizontally at least **12 inches** (305) beyond the **top and bottom** risers.
4) There is not enough **headroom (80"or 2032)** for the doors under the landing or stairs.
5) Stairs block the egress door or the lobby door on the ground floor.
6) Forgetting to show the 30" x48" area of refuge, or the 24" and 5'-0" x 5'-0" clearance space at the pull side of the egress door. Forget stair width increase to **48" (1219) clear between handrails** when area of refuge is required. This can happen if you do NOT read ALL the NCARB requirements.
7) The width of a landing is less than the width of the stair.
8) Designing two stairs instead of ONE stair as required by the NCARB program.
9) Riser height is not the same throughout the stair runs.
10) Number of risers and riser heights do NOT match or add up to the elevations for the storage level or the second floor level.
11) When the cut stair tool is used, the data for the cut stair on the second floor does NOT match the data for the same cut stair on the first floor, such as number of risers and treads, elevations for the top and bottom of the cut stair.

4. A summary of the critical dimensions

1) The **minimum** dimension for the **landing:**   5'-0" (1524)
2) **Stair width (4'-8" minimum):**   5'-0" (1524)
3) **Treads:**   11" (279)
4) **Risers:**   7" (178)
5) Handrail **extension** at **top and bottom** of the stair runs: 12" (305)
6) **Clearance** for the **pull side** of the egress door: 24" and 5'-0" x 5'-0" (610 and 1524 x 1524)
7) **Refuge area:**   30"x48" (762 x 1219)
8) **Headroom:**   80" (2032)
9) Ground level exit **door height**   8'-0" (2438)
10) **Clearance** under the intermediate landing   7'-2" > 80" (2184 > 2032)

(Adequate to clear the headroom and the ground level exit door below)

D. Mock Exam Solution: BDCS Roof Plan Vignette

1. Major criteria, overall strategy and tips for the roof plan vignette:

1) The building has one high roof and one low roof.
2) Outside edges of the roof planes must coincide with the dashed lines indicating the outermost edges of the roofs. Gutters and downspouts can be placed beyond with the dashed lines.
3) The slope for the roof over the Multi-purpose Room and Lobby shall be between 6:12 and 12:12.
4) The slope for the roof over the remaining spaces shall be between 2:12 and 5:12.
5) Check the elevations of the low points of upper roof and high points of the lower roof, and make sure you leave adequate space to accommodate the 1'-6" (457) thick roof and structural assembly and the **2'-6" (762) high clerestory window** in the west wall.
6) Provide skylights for rooms have no windows and no clerestory window.
7) Halls, storage rooms, or closets do not need skylights.
8) Check the elevations of the low points of the lower roof, and make sure you leave adequate space to accommodate the 1'-6" (457) thick roof and structural assembly and the **8'-0" high first floor ceiling**.
9) Do NOT miss the gutters or downspouts.
10) Do NOT miss the cricket for the chimney.
11) Do NOT miss the plumbing vent stacks, and exhaust fan vents for the restrooms and the Break Room.
12) Do NOT miss the skylights at the Men's Restroom and the Women's Restroom.
13) Place the HVAC condensing unit on a roof with a slope of 5:12 or less, but NOT in front of the clerestory window.
14) Make sure the HVAC condensing unit has the required 3'-0" (914) minimum clearance from all roof edges.

2. Step-by-step solution for the mock exam roof plan vignette using the official NCARB BDCS practice program:

1) Open the NCARB graphic program, and find a large blank area.
2) Since we have all the critical dimensions for the buildings, we can use the **Sketch** tool of NCARB graphic program to draw a background outline for the major spaces and the building that we will work on (Figure 4.30).

3) **Calculating the roof heights:**
The elevation for the lowest points of lower roof = the roof and structural assembly thickness + the first floor ceiling height = (1'-6") + (8'-0") = 9'-6" (2895)

The **minimum** difference between the upper roof and lower roof = the roof and structural assembly thickness + the clerestory window height = (1'-6") + (2'-6") = 4'-0" (1219)

4) This is a symmetrical building, we are going to use a ridge to separate the lower roof into two roof planes, and use a ridge to separate the upper roof into two roof planes.

Use **Draw > Roof Plane** to draw part of the lower roof (Figure 4.31).

Note: This is an important step for this solution. Pay attention the shape of the roof plane. This roof area is covered by **one** *roof plane instead of* **multiple** *roof planes because it has the same slope.*

5) Click on **Set roof**, and then click on the **arrow symbol** to set the direction of the slope. Click on a **question mark** next to the arrow to bring up a dialogue box, use the **up or down arrows** to set the slope ratio to 2:12. Click on **Set roof**, and then click on the other **question mark** next to the elevation marker to bring up a dialogue box, use the up or down arrows to set the elevation for the lowest points of lower roof to 9'-6" or 2895 (Figure 4.32).

Note: After you bring up a dialogue box, the **up or down arrow** *is* **tricky and hard to use**. *For example, if you want to set an elevation to 9'-6", if you set the 9' elevation, and then try to set the 6", you may NOT be able to do it, you'll get 9'-5" or 9'-7", but you may NOT get 9'-6". You may have to set it to 8'-9" first, and then keep clicking on the* **up arrow next to the inches** *to increase it as 8'-9", 8'-11", 9'-0", 9'-1" and so on until you reach 9'-6". You need to practice and play with the NCARB software and beware of tricky things like this. Otherwise, you will be wasting your valuable time to explore this in the real ARE test.*

6) Draw the remaining lower roof plane using commands similar to steps 4) and 5) (Figure 4.33).

Because we are using 2:12 or 1:6 slope, and the horizontal distance between the low edge of the lower roof and the upper left-hand corner of the admin area is (12'-5") + (10'-7") = 23' (3785 + 3225 = 7010), so their difference between their elevations is 23'/6 = 3'-10" (7010/6 = 1168)

The elevation of the lower roof at the upper left-hand corner of the admin area = (9'-6") + (3'-10") = 13'-4" (2896 + 1168 = 4064)

7) Use **Zoom** to zoom in every corner of the roof plans, and use **Move, Adjust to** adjust the edge of the roof planes to align with the dashed lines if necessary.

8) Draw the upper roof using commands similar to steps 4) and 5) (Figure 4.34).

We set the ridgeline in the middle of the upper roof.

The elevation for the lowest points of upper roof = the elevation of the lower roof next to the low edge of the higher roof + (4'-0") = (13'-4") + (4'-0") = 17'-4" (4064 + 1219 = 5283)

The slope of the upper roof is 6:12 or 1:2.

The elevation for the ridgeline of the upper roof = (17'-4") + (10'/2) = 22'-4" (5283 + 3048/2 = 6807)

9) Use **Zoom** to zoom into every corner of the roof plans, and use **Move, Adjust to** adjust the edge of the roof planes to align with the lower roof planes if necessary.

10) Use **Draw > Clerestory** to draw the clerestory window in the east wall below the upper roof.

Note: Because of the limitations of the NCARB software, the width of the clerestory window is set to the original NCARB vignette program, and may be larger than we need for this solution. There is nothing we can do about it, but you get the intent.

11) Use **Draw > HVAC condenser** to draw the HVAC condensing unit above the corridor on the lower roof (Figure 4.35).

Make sure the HVAC condensing unit has the required **3'-0" (914) minimum clearance** from all roof edges.

12) Because the Break Room and the Women's Restroom have no window or clerestory window, each needs skylight(s).

Use **Draw > Skylight** to draw two skylights for the Break Room and a skylight for the Women's Restroom. Use **Rotate** to rotate each of the skylight 90 degrees.

13) Use **Draw > Exhaust Fan Vent** to draw an exhaust fan for the Men's Restroom, an exhaust fan for the Women's Restroom, and an exhaust fan near the sink in the Break Room.

14) Use **Draw > Plumbing Vent Stack** to draw a plumbing vent stack inside the common walls between the Men's Restroom and the Women's Restroom, and the wall behind the Break Room sink (Figure 4.36).

15) There is no need to add cricket for the chimney because the water drains away from the chimney.

16) Use **Draw > Gutter** to draw a gutter along the **low** edge of **every** slope roof plan.

*Note: There is a defect with the gutter command of the NCARB software. Once you place it, it is almost impossible to modify it. There are two ways to solve this problem: One way is **Zoom out** first, and then click on the gutter a few times to select and **erase** it, and then **redraw** a new one. The other way is to select the gutter while clicking on the side that you drew it with (usually **the side closest to the building**). You can then move the gutter like* *any other element. If you want to adjust the gutter length, you need to click on the* **intersection** *at the end of the gutter and on* **the side closest to the building**, *and then adjust it.*

17) Use **Draw > Downspout** to draw a downspout on both end of **each** gutter (Figure 4.37).

18) Use **Zoom** to zoom in every corner of the roof plans, and use **Move, Adjust to** adjust the gutters and downspouts if necessary.

19) Use **Draw > Flashing** to add flashings to **ALL** intersections at walls, roofs and chimney. This is the final solution (Figure 4.38).

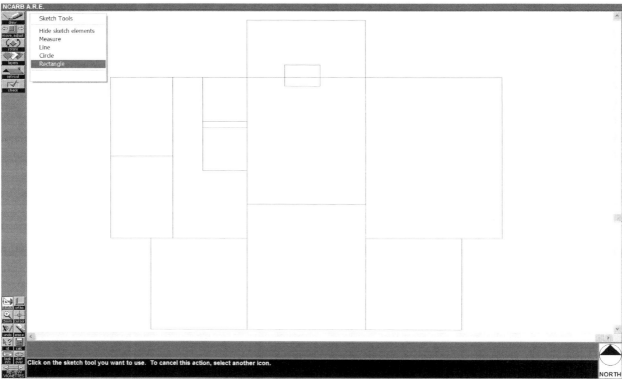

Figure 4.30   Use the **Sketch** tool of NCARB graphic program to draw a background outline for the major spaces and the building that we will work on.

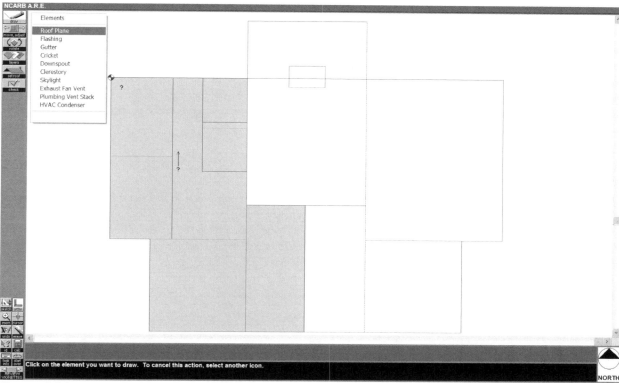

Figure 4.31   Use **Draw > Roof Plan** to draw part of the lower roof.

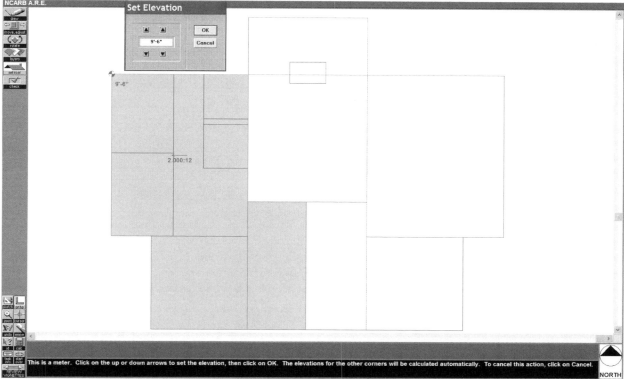

Figure 4.32   Click on **Set roof**, and then click on the **arrow symbol** to set the direction of the slope. Click on the question marks to set the **slope ratio**, and the **elevation**.

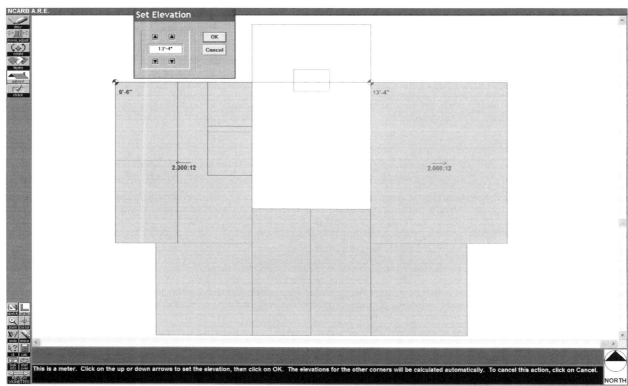

Figure 4.33   Draw the remaining lower roof plane using commands similar to steps 4) and 5).

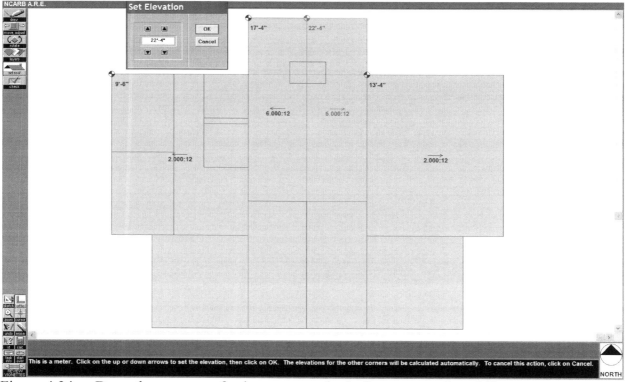

Figure 4.34   Draw the upper roof using commands similar to steps 4) and 5).

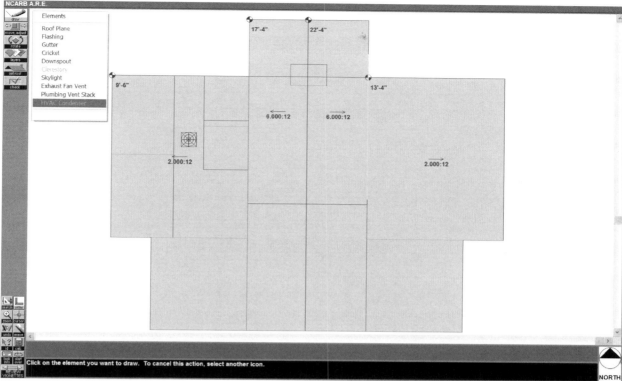

Figure 4.35   Use **Draw > Clerestory** to draw the clerestory. Use **Draw > HVAC condenser** to draw the HVAC condensing unit above the corridor on the lower roof.

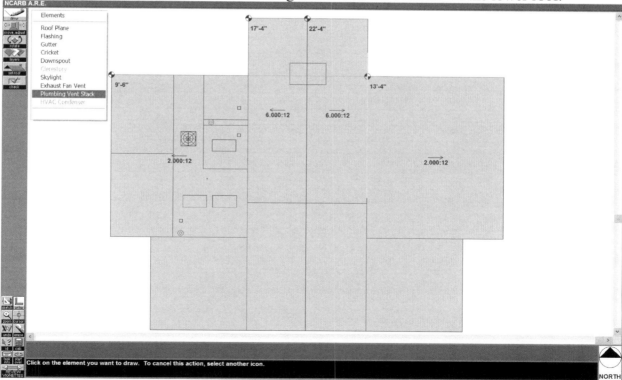

Figure 4.36   Adding **skylights**, **exhaust fan vent**, and **plumbing vent stacks**.

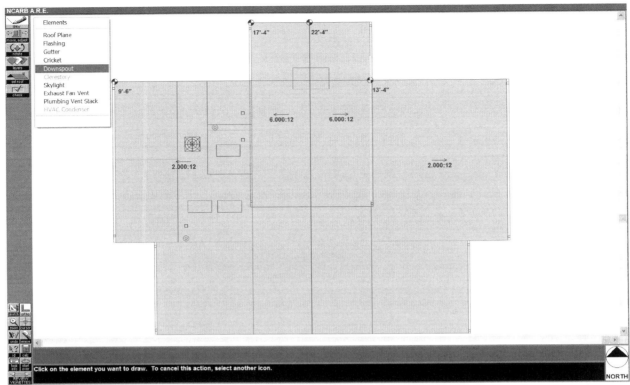

Figure 4.37   Use **Draw > Downspout** to draw a downspout on both end of **each** gutter.

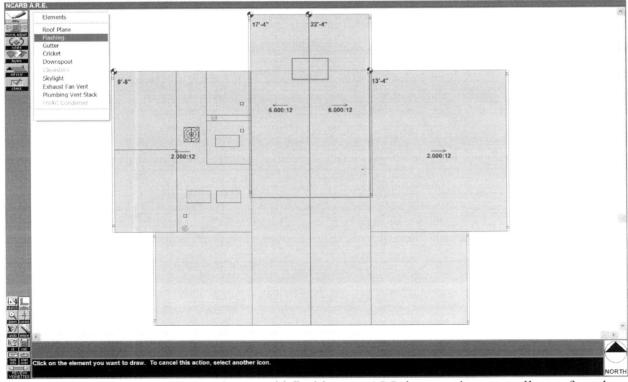

Figure 4.38   Use **Draw > Flashing** to add flashings to **ALL** intersections at walls, roofs and chimney. This is the final solution.

3. Notes on mock exam traps

Several **common errors** or **traps** into which you may fall:

1) Draw part of the roof planes outside of the **dashed lines** (or building outline in this case):
   - Outside edges of the roof planes must coincide with the dashed lines indicating the outermost edges of the roofs. Gutters and downspouts can be placed beyond the dashed lines.

2) Show the **wrong slop ratio**:
   - The slope for the roof over the Multi-purpose Room shall be between 6:12 and 12:12.
   - The slope for the roof over the remaining spaces shall be between 2:12 and 5:12.

3) Not enough **clearance space** for the roof and structural assembly and **clerestory window:**
   - Check the elevations of the low points of upper roof and the lower roof below at the same location, and make sure you leave adequate space to accommodate the 1'-6" (457) thick roof and structural assembly and the **2'-6" (762) high clerestory window** in the east wall.

4) Miss the **skylights**:
   - Provide skylights for rooms have no windows and no clerestory window.
   - Halls, storage rooms, or closets do not need skylights.

5) Not enough **clearance space** for first floor ceiling:
   - Check the elevations of the low points of the lower roof, and make sure you leave adequate space to accommodate the 1'-6" thick roof and structural assembly and the **8'-0" high first floor ceiling**.

6) Miss the gutters, downspouts, the plumbing vent stacks, or the exhaust fan vents for the restrooms and the kitchen.

7) Not enough **clearance space** clearance for the HVAC condensing unit:
   - Place the HVAC condensing unit on a roof with a slope of 5:12 or less, but NOT in front of the clerestory window.
   - Make sure the HVAC condensing unit has the required 3'-0" (914) minimum clearance from all roof edges.

4. A summary of the critical dimensions

1) First floor **ceiling** height: 8'-0" (2438)
2) Roof and structural assembly thickness: 1'-6" (457)
3) Low roof:
   - Slope ratio: 2:12
   - Elevation of the **lowest** point: 9'-6" (2896)
   - Elevation of the **lower** roof at clerestory window: 13'-4" (4064)
   - Clearance under **lower** roof: 8'-0" (2438)

4) Upper roof:
   - Slope ratio: 6:12
   - Elevation of the **highest** point or the **ridge line**:   22'-4" (6807)
   - Elevation of the **lowest** point at clerestory window: 17'-4" (5283)
   - Clearance under **upper** roof for clerestory window and roof structure: (2'-6") + (1'-6") = 4'-0"
                                                               (762 + 457 = 1219)

Appendixes

A. List of Figures

B. Official reference materials suggested by NCARB

1. General NCARB reference materials for ARE:

Per NCARB, all candidates should become familiar with the latest version of the following codes:

International Code Council, Inc. (ICC, 2006)
International Building Code
International Mechanical Code
International Plumbing Code

National Fire Protection Association (NFPA)
Life Safety Code (NFPA 101)
National Electrical Code (NFPA 70)

National Research Council of Canada
National Building Code of Canada
National Plumbing Code of Canada
National Fire Code of Canada

American Institute of Architects
AIA Documents - 2007

Candidates should be familiar with the Standard on Accessible and Usable Buildings and Facilities (ICC/ANSI A117.1-98)

2. Official NCARB reference materials for the Building Design and Construction Systems (BDCS) division:

The Architect's Handbook of Professional Practice
Joseph A. Demkin, AIA, Executive Editor
The American Institute of Architects
JohnWiley & Sons, latest edition
A comprehensive book covers all aspect of architectural practice, including 2 CDs containing the sample AIA contract documents.

The Architect's Portable Handbook
Second Edition
Pat Guthrie
McGraw-Hill, 2003

The Architect's Studio Companion: Technical Guidelines for Preliminary Design
Edward Allen and Joseph Iano
JohnWiley & Sons, latest edition

Architectural Graphic Standards
Charles G. Ramsey and Harold R. Sleeper
The American Institute of Architects
JohnWiley & Sons, latest edition

Building Codes Illustrated: A Guide to Understanding the International Building Code
Second Edition
Francis D. K. Ching and Steven R.Winkel, FAIA
JohnWiley & Sons, 2007
A valuable interpretive guide with many useful line drawings. A great timesaver.

Building Construction Illustrated
Francis D. K. Ching and Cassandra Adams
JohnWiley & Sons, latest edition

Building Design and Construction Handbook,
Sixth Edition
Frederick S. Merritt and Jonathan T. Ricketts
McGraw-Hill, 2000

Building Construction Illustrated
Francis D. K. Ching and Cassandra Adams
John Wiley & Sons, latest edition

Building Design and Construction Handbook
Sixth Edition
Frederick S. Merritt and Jonathan T. Ricketts
McGraw-Hill, 2000

Dictionary of Architecture and Construction
Cyril M. Harris, Editor
McGraw-Hill, 2005

Fundamentals of Building Construction, Materials, and Methods
Fourth Edition
Edward Allen
John Wiley & Sons, latest edition

Historic Preservation: An Introduction to Its History, Principles, and Practice
Norman Tyler
W.W. Norton & Company, latest edition

The HOK Guidebook to Sustainable Design
Sandra F. Mendler, AIA, and William Odell, AIA
John Wiley & Sons, 2006

Illustrated Dictionary of Architecture
Second Edition
Ernest Burden
McGraw-Hill, 2002

Interior Graphic Standards
Maryrose McGowan and Kelsey Kruse
John Wiley & Sons, 2003

Landscape Planning: Environmental Applications
Fourth Edition
William M.Marsh
John Wiley & Sons, 2005

Means Building Construction Cost Data
RS Means Company, latest edition

Sun, Wind, and Light: Architectural Design Strategies
Second Edition
G. Z. Brown and Mark DeKay
John Wiley & Sons, 2000

Time-Saver Standards for Architectural Design Data
Donald Watson, Michael J. Crosbie, and
John Hancock Callender, Editors
McGraw-Hill, latest edition

Time-Saver Standards for Building Materials & Systems: Design Criteria and Selection Data
Donald Watson, Editor
McGraw-Hill, latest edition

Time-Saver Standards for Building Types
Fourth Edition
Joseph De Chiara and Michael J. Crosbie, Editors
McGraw-Hill, latest edition

Time-Saver Standards for Interior Design and Space Planning
Joseph De Chiara, Editor, Julius Panero, Editor, and Martin Zelnik
McGraw-Hill, latest edition

C. Other reference materials

Access Board, *ADAAG Manual: A Guide to the American with Disabilities Accessibility Guidelines*. East Providence, RI: BNI Building News. ADA Standards for Accessible Design are available as FREE PDF files at

http://www.ada.gov/

AND
http://www.access-board.gov/adaag/html/figures/index.html

Chen, Gang. *Building Construction: Project Management, Construction Administration, Drawings, Specs, Detailing Tips, Schedules, Checklists, and Secrets Others Don't Tell You (Architectural Practice Simplified, 2nd edition)*. ArchiteG, Inc.
A good introduction to the architectural practice and construction documents and service. It includes discussions of MasterSpec format and specification sections.

Chen, Gang. *LEED GA Exam Guide: A Must-Have for the LEED Green Associate Exam: Comprehensive Study Materials, Sample Questions, Mock Exam, Green Building LEED Certification, and Sustainability*. ArchiteG, Inc., latest edition.
A good introduction to green buildings and the LEED building rating system.

Ching, Francis. *Architecture: Form, Space, & Order*. Wiley, latest edition.
It is one of the best architectural books that you can have. I still flip through it every now and then. It is a great book for inspiration.

Frampton, Kenneth. *Modern Architecture: A Critical History*. Thames and Hudson, London, latest edition.
A valuable resource for architectural history.

Jarzombek, Mark M. (Author), Vikramaditya Prakash (Author), Francis D. K. Ching (Editor). *A Global History of Architecture*. Wiley, latest edition.
A valuable and comprehensive resource for architectural history with 1000 b & w photos, 50 color photos, and 1500 b & w illustrations. It doesn't limit the topic on a Western perspective, but rather through a global vision.

Trachtenberg, Marvin and Isabelle Hyman. *Architecture: From Pre-history to Post-Modernism*. Prentice Hall, Englewood Cliffs, NJ latest edition.
A valuable and comprehensive resource for architectural history.

D. Definition of Architects and Some Important Information about Architects and the Profession of Architecture

Architects, Except Landscape and Naval
- Nature of the Work
- Training, Other Qualifications, and Advancement
- Employment
- Job Outlook
- Projections Data
- Earnings
- OES Data
- Related Occupations
- Sources of Additional Information

Significant Points
- About 1 in 5 architects are self-employed—more than 2 times the proportion for all occupations.
- Licensing requirements include a professional degree in architecture, at least 3 years of practical work training, and passing all divisions of the Architect Registration Examination.
- Architecture graduates may face competition, especially for jobs in the most prestigious firms.

Nature of the Work
People need places in which to live, work, play, learn, worship, meet, govern, shop, and eat. These places may be private or public; indoors or out; rooms, buildings, or complexes, and architects design them. Architects are licensed professionals trained in the art and science of building design who develop the concepts for structures and turn those concepts into images and plans.

Architects create the overall aesthetic and look of buildings and other structures, but the design of a building involves far more than its appearance. Buildings also must be functional, safe, and economical and must suit the needs of the people who use them. Architects consider all these factors when they design buildings and other structures.

Architects may be involved in all phases of a construction project, from the initial discussion with the client through the entire construction process. Their duties require specific skills—designing, engineering, managing, supervising, and communicating with clients and builders. Architects spend a great deal of time explaining their ideas to clients, construction contractors, and others. Successful architects must be able to communicate their unique vision persuasively.

The architect and client discuss the objectives, requirements, and budget of a project. In some cases, architects provide various pre-design services: conducting feasibility and environmental impact studies, selecting a site, preparing cost analysis and land-use studies, or specifying the requirements the design must meet. For example, they may determine space requirements by researching the numbers and types of potential users of a building. The architect then prepares drawings and a report presenting ideas for the client to review.

After discussing and agreeing on the initial proposal, architects develop final construction plans that show the building's appearance and details for its construction. Accompanying these plans are drawings of the structural system; air-conditioning, heating, and ventilating systems; electrical systems; communications systems; plumbing; and, possibly, site and landscape plans. The plans also specify the building materials and, in some cases, the interior furnishings. In developing designs, architects follow building codes, zoning laws, fire regulations, and other ordinances, such as those requiring easy access by people who are disabled. Computer-aided design and drafting (CADD) and Building Information Modeling (BIM) technology has replaced traditional paper and pencil as the most common method for creating design and construction drawings. Continual revision of plans on the basis of client needs and budget constraints is often necessary.

Architects may also assist clients in obtaining construction bids, selecting contractors, and negotiating construction contracts. As construction proceeds, they may visit building sites to make sure that contractors follow the design, adhere to the schedule, use the specified materials, and meet work quality standards. The job is not complete until all construction is finished, required tests are conducted, and construction costs are paid. Sometimes, architects also provide post-construction services, such as facilities management. They advise on energy efficiency measures, evaluate how well the building design adapts to the needs of occupants, and make necessary improvements.

Often working with engineers, urban planners, interior designers, landscape architects, and other professionals, architects in fact spend a great deal of their time coordinating information from, and the work of, other professionals engaged in the same project.

They design a wide variety of buildings, such as office and apartment buildings, schools, churches, factories, hospitals, houses, and airport terminals. They also design complexes such as urban centers, college campuses, industrial parks, and entire communities.

Architects sometimes specialize in one phase of work. Some specialize in the design of one type of building—for example, hospitals, schools, or housing. Others focus on planning and pre-design services or construction management and do minimal design work.

Work environment. Usually working in a comfortable environment, architects spend most of their time in offices consulting with clients, developing reports and drawings, and working with other architects and engineers. However, they often visit construction sites to review the progress of projects. Although most architects work approximately 40 hours per week, they often have to work nights and weekends to meet deadlines.

Training, Other Qualifications, and Advancement

There are three main steps in becoming an architect. First is the attainment of a professional degree in architecture. Second is work experience through an internship, and third is licensure through the passing of the Architect Registration Exam.

Education and training. In most States, the professional degree in architecture must be from one of the 114 schools of architecture that have degree programs accredited by the National Architectural Accrediting Board. However, State architectural registration boards set their own standards, so graduation from a non-accredited program may meet the educational requirement for licensing in a few States.

Three types of professional degrees in architecture are available: a 5-year bachelor's degree, which is most common and is intended for students with no previous architectural training; a 2-year master's degree for students with an undergraduate degree in architecture or a related area; and a 3- or 4-year master's degree for students with a degree in another discipline.

The choice of degree depends on preference and educational background. Prospective architecture students should consider the options before committing to a program. For example, although the 5-year bachelor of architecture offers the fastest route to the professional degree, courses are specialized, and if the student does not complete the program, transferring to a program in another discipline may be difficult. A typical program includes courses in architectural history and theory, building design with an emphasis on CADD, structures, technology, construction methods, professional practice, math, physical sciences, and liberal arts. Central to most architectural programs is the design studio, where students apply the skills and concepts learned in the classroom, creating drawings and three-dimensional models of their designs.

Many schools of architecture also offer post-professional degrees for those who already have a bachelor's or master's degree in architecture or other areas. Although graduate education beyond the professional degree is not required for practicing architects, it may be required for research, teaching, and certain specialties.

All State architectural registration boards require architecture graduates to complete a training period—usually at least 3 years—before they may sit for the licensing exam. Every State, with the exception of Arizona, has adopted the training standards established by the Intern Development Program, a branch of the American Institute of Architects and the National Council of Architectural Registration Boards (NCARB). These standards stipulate broad training under the supervision of a licensed architect. Most new graduates complete their training period by working as interns at architectural firms. Some States allow a portion of the training to occur in the offices of related professionals, such as engineers or general contractors. Architecture students who complete internships while still in school can count some of that time toward the 3-year training period.

Interns in architectural firms may assist in the design of one part of a project, help prepare architectural documents or drawings, build models, or prepare construction drawings on CADD. Interns also may research building codes and materials or write specifications for building materials, installation criteria, the quality of finishes, and other, related details.

Licensure. All States and the District of Columbia require individuals to be licensed (registered) before they may call themselves architects and contract to provide architectural services. During the time between graduation and becoming licensed, architecture school graduates generally work in the field under the supervision of a licensed architect who takes legal responsibility for all work. Licensing requirements include a professional degree in architecture, a period of practical training or internship, and a passing score on all divisions of the Architect Registration Examination. The examination is broken into nine divisions consisting of either multiple choice or graphical questions. The eligibility period for completion of all divisions of the exam varies by State.

Most States also require some form of continuing education to maintain a license, and many others are expected to adopt mandatory continuing education. Requirements vary by State but usually involve the completion of a certain number of credits annually or biennially through workshops, formal university classes, conferences, self-study courses, or other sources.

Other qualifications. Architects must be able to communicate their ideas visually to their clients. Artistic and drawing ability is helpful, but not essential, to such communication. More important are a visual orientation and the ability to understand spatial relationships. Other important qualities for anyone interested in becoming an architect are creativity and the ability to work independently and as part of a team. Computer skills are also required for writing specifications, for 2- and 3- dimensional drafting using CADD programs, and for financial management.

Certification and advancement. A growing number of architects voluntarily seek certification by the National Council of Architectural Registration Boards. Certification is awarded after independent verification of the candidate's educational transcripts, employment record, and professional references. Certification can make it easier to become licensed across States. In fact, it is the primary requirement for reciprocity of licensing among State Boards that are NCARB members. In 2007, approximately one-third of all licensed architects had this certification.

After becoming licensed and gaining experience, architects take on increasingly responsible duties, eventually managing entire projects. In large firms, architects may advance to supervisory or managerial positions. Some architects become partners in established firms, while others set up their own practices. Some graduates with degrees in architecture also enter related fields, such as graphic, interior, or industrial design; urban planning; real estate development; civil engineering; and construction management.

Employment

Architects held about 132,000 jobs in 2006. Approximately 7 out of 10 jobs were in the architectural, engineering, and related services industry—mostly in architectural firms with fewer than five workers. A small number worked for residential and nonresidential building construction firms and for government agencies responsible for housing, community planning, or construction of government buildings, such as the U.S.

Departments of Defense and Interior, and the General Services Administration. About 1 in 5 architects are self-employed.

Job Outlook

Employment of architects is expected to grow faster than the average for all occupations through 2016. Keen competition is expected for positions at the most prestigious firms, and opportunities will be best for those architects who are able to distinguish themselves with their creativity.

Employment change. Employment of architects is expected to grow by 18 percent between 2006 and 2016, which is <u>faster than the average</u> for all occupations. Employment of architects is strongly tied to the activity of the construction industry. Strong growth is expected to come from nonresidential construction as demand for commercial space increases. Residential construction, buoyed by low interest rates, is also expected to grow as more people become homeowners. If interest rates rise significantly, home building may fall off, but residential construction makes up only a small part of architects' work.

Current demographic trends also support an increase in demand for architects. As the population of Sunbelt States continues to grow, the people living there will need new places to live and work. As the population continues to live longer and baby-boomers begin to retire, there will be a need for more healthcare facilities, nursing homes, and retirement communities. In education, buildings at all levels are getting older and class sizes are getting larger. This will require many school districts and universities to build new facilities and renovate existing ones.

In recent years, some architecture firms have outsourced the drafting of construction documents and basic design for large-scale commercial and residential projects to architecture firms overseas. This trend is expected to continue and may have a negative impact on employment growth for lower level architects and interns who would normally gain experience by producing these drawings.

Job prospects. Besides employment growth, additional job openings will arise from the need to replace the many architects who are nearing retirement, and others who transfer to other occupations or stop working for other reasons. Internship opportunities for new architectural students are expected to be good over the next decade, but more students are graduating with architectural degrees and some competition for entry-level jobs can be anticipated. Competition will be especially keen for jobs at the most prestigious architectural firms as prospective architects try to build their reputation. Prospective architects who have had internships while in school will have an advantage in obtaining intern positions after graduation. Opportunities will be best for those architects that are able to distinguish themselves from others with their creativity.

Prospects will also be favorable for architects with knowledge of "green" design. Green design, also known as sustainable design, emphasizes energy efficiency, renewable resources such as energy and water, waste reduction, and environmentally friendly design, specifications, and materials. Rising energy costs and increased concern about the environment has led to many new buildings being built green.

Some types of construction are sensitive to cyclical changes in the economy. Architects seeking design projects for office and retail construction will face especially strong competition for jobs or clients during recessions, and layoffs may ensue in less successful firms. Those involved in the design of institutional buildings, such as schools, hospitals, nursing homes, and correctional facilities, will be less affected by fluctuations in the economy. Residential construction makes up a small portion of work for architects, so major changes in the housing market would not be as significant as fluctuations in the nonresidential market.

Despite good overall job opportunities, some architects may not fare as well as others. The profession is geographically sensitive, and some parts of the Nation may have fewer new building projects. Also, many firms specialize in specific buildings, such as hospitals or office towers, and demand for these buildings may vary by region. Architects may find it increasingly necessary to gain reciprocity in order to compete for the best jobs and projects in other States.

Projections Data

Projections data from the National Employment Matrix

Occupational title	SOC Code	Employment, 2006	Projected employment, 2016	Change, 2006-16		Detailed statistics
				Number	Percent	
Architects, except landscape and naval	17-1011	132,000	155,000	23,000	18	PDF zipped XLS

   NOTE: Data in this table are rounded. See the discussion of the employment projections table in the *Handbook* introductory chapter on *Occupational Information Included in the Handbook*.

Earnings
Median annual earnings of wage-and-salary architects were $64,150 in May 2006. The middle 50 percent earned between $49,780 and $83,450. The lowest 10 percent earned less than $39,420, and the highest 10 percent earned more than $104,970. Those just starting their internships can expect to earn considerably less.

Earnings of partners in established architectural firms may fluctuate because of changing business conditions. Some architects may have difficulty establishing their own practices and may go through a period when their expenses are greater than their income, requiring substantial financial resources.

Many firms pay tuition and fees toward continuing education requirements for their employees.

For the latest wage information:
The above wage data are from the Occupational Employment Statistics (OES) survey program, unless otherwise noted. For the latest National, State, and local earnings data, visit the following pages:
Architects, except landscape and naval

Related Occupations
Architects design buildings and related structures. Construction managers, like architects, also plan and coordinate activities concerned with the construction and maintenance of buildings and facilities. Others who engage in similar work are landscape architects, civil engineers, urban and regional planners, and designers, including interior designers, commercial and industrial designers, and graphic designers.

Sources of Additional Information

Disclaimer:
Links to non-BLS Internet sites are provided for your convenience and do not constitute an endorsement.

Information about education and careers in architecture can be obtained from:
- The American Institute of Architects, 1735 New York Ave. NW., Washington, DC 20006. Internet: http://www.aia.org
- Intern Development Program, National Council of Architectural Registration Boards, Suite 1100K, 1801 K St. NW., Washington, D.C. 20006. Internet: http://www.ncarb.org OOH ONET Codes 17-1011.00"

Quoted from: Bureau of Labor Statistics, U.S. Department of Labor, Occupational Outlook Handbook, 2008-09 Edition, Architects, Except Landscape and Naval, on the Internet at **http://www.bls.gov/oco/ocos038.htm** (visited November 30, 2008).
Last Modified Date: December 18, 2007

Note: Please check the website above for the latest information.

E. AIA Compensation Survey

Every 3 years, AIA publishes a Compensation Survey for various positions at architectural firms across the country. It is a good idea to find out the salary before you make the final decision to become an architect. If you are already an architect, it is also a good idea to determine if you are underpaid or overpaid.

See following link for some sample pages for the 2008 AIA Compensation Survey:

http://www.aia.org/aiaucmp/groups/ek_public/documents/pdf/aiap072881.pdf

F. So ... You would Like to Study Architecture

To study architecture, you need to learn how to draft, how to understand and organize spaces and the interactions between interior and exterior spaces, how to do design, and how to communicate effectively. You also need to understand the history of architecture.

As an architect, a leader for a team of various design professionals, you not only need to know architecture, but also need to understand enough of your consultants' work to be able to coordinate them. Your consultants include soils and civil engineers, landscape architects, structural, electrical, mechanical, and plumbing engineers, interior designers, sign consultants, etc.

There are two major career paths for you in architecture: practice as an architect or teach in colleges or universities. The earlier you determine which path you are going to take, the more likely you will be successful at an early age. Some famous and well-respected architects, like my USC alumnus Frank Gehry, have combined the two paths successfully. They teach at the universities and have their own architectural practice. Even as a college or university professor, people respect you more if you have actual working experience and have some built projects. If you only teach in colleges or universities but have no actual working experience and have no built projects, people will consider you as a "paper" architect, and they are not likely to take you seriously, because they will think you probably do not know how to put a real building together.

In the U.S., if you want to practice architecture, you need to obtain an architect's license. It requires a combination of passing scores on the Architectural Registration Exam (ARE) and 8 years of education and/or qualified working experience, including at least 1 year of working experience in the U.S. Your working experience needs to be under the supervision of a licensed architect to be counted as qualified working experience for your architect's license.

If you work for a landscape architect or civil engineer or structural engineer, some states' architectural licensing boards will count your experience at a discounted rate for the qualification of your architect's license. For example, 2 years of experience working for a civil engineer may be counted as 1 year of qualified experience for your architect's license. You need to contact your state's architectural licensing board for specific licensing requirements for your state.

If you want to teach in colleges or universities, you probably want to obtain a master's degree or a Ph.D. It is not very common for people in the architectural field to have a Ph.D. One reason is that there are few Ph.D. programs for architecture. Another reason is that architecture is considered a profession and requires a license. Many people think an architect's license is more important than a Ph.D. degree. In many states, you need to have an architect's license to even use the title "architect," or the terms "architectural" or "architecture" to advertise your service. You cannot call yourself an architect if you do not have an architect's license, even if you have a Ph.D. in architecture. Violation of these rules brings punishment.

To become a tenured professor, you need to have a certain number of publications and pass the evaluation for the tenure position. Publications are very important for tenure track positions. Some people say for the tenured track positions in universities and colleges, it is "publish or perish."

The American Institute of Architects (AIA) is the national organization for the architectural profession. Membership is voluntary. There are different levels of AIA membership. Only licensed architects can be (full) AIA members. If you are an architectural student or an intern but not a licensed architect yet, you can join as an associate AIA member. Contact AIA for detailed information.

The National Council of Architectural Registration Boards (NCARB) is a nonprofit federation of architectural licensing boards. It has some very useful programs, such as IDP, to assist you in obtaining your architect's license. Contact NCARB for detailed information.

Back Page Promotion

You may be interested in some other books written by Gang Chen:

A. **ARE Mock exam series.** See following link:
   http://www.GreenExamEducation.com

B. **LEED Exam Guides series.** See following link:
   http://www.GreenExamEducation.com

C. ***Building Construction:*** *Project Management, Construction Administration, Drawings, Specs, Detailing Tips, Schedules, Checklists, and Secrets Others Don't Tell You (Architectural Practice Simplified, 2nd edition)*
   http://www.ArchiteG.com

D. ***Planting Design Illustrated***
   http://outskirtspress.com/agent.php?key=11011&page=GangChen

ARE Mock Exam Series

CONSTRUCTION DOCUMENTS AND SERVICE (CDS) ARE MOCK EXAM (ARCHITECT REGISTRATION EXAM)*: ARE OVERVIEW, EXAM PREP TIPS, MULTIPLE-CHOICE QUESTIONS AND GRAPHIC VIGNETTES, SOLUTIONS AND EXPLANATIONS* (Published May 22, 2011)

A PRACTICAL GUIDE FOR THE CONSTRUCTION DOCUMENTS & SERVICE DIVISION OF THE ARE

To become a licensed architect, you need to have a proper combination of education and/or experience, meeting your Board of Architecture's special requirements, as well as passing all seven divisions of the Architect Registration Examinations (ARE).

This book provides an ARE exam overview, resources, exam prep and exam taking techniques, tips and guide, a realistic and complete set of Mock Exams, solutions, and explanations for the Construction Documents & Service (CDS) Division of the ARE.

This book covers the following subjects:
1. ARE, IDP, and Education Requirements
2. ARE Exam Content, Format, and Prep strategies
3. Bidding Procedures and Documents
4. Codes and Regulations
5. Environmental Issues
6. Construction Administration Services
7. Construction Drawings
8. Project Manual and Specifications
9. Contractual Documents
10. Project and Practice Management
11. Two Building Section Vignettes with Step-By-Step Solutions Using NCARB Practice Program Software

This book includes 100 challenging questions in the same difficulty level and format as the real exam (multiple-choice, check-all-that-apply, and fill-in-the-blank), and two graphic vignettes. It will help you pass the CDS division of the ARE and become a licensed architect!

LEED Exam Guides Series*: Comprehensive Study Materials, Sample Questions, Mock Exam, Building LEED Certification and Going Green*

LEED (Leadership in Energy and Environmental Design) is the most important trend of development, and it is revolutionizing the construction industry. It has gained tremendous momentum and has a profound impact on our environment.

From LEED Exam Guides series, you will learn how to

1. Pass the LEED Green Associate Exam and various LEED AP + exams (each book will help you with a specific LEED exam).

2. Register and certify a building for LEED certification.

3. Understand the intent for each LEED prerequisite and credit.

4. Calculate points for a LEED credit.

5. Identify the responsible party for each prerequisite and credit.

6. Earn extra credit (exemplary performance) for LEED.

7. Implement the local codes and building standards for prerequisites and credit.

8. Receive points for categories not yet clearly defined by USGBC.

There is currently NO official book on the LEED Green Associate Exam, and most of the existing books on LEED and LEED AP are too expensive and too complicated to be practical and helpful. The pocket guides in LEED Exam Guides series fill in the blanks, demystify LEED, and uncover the tips, codes, and jargon for LEED as well as the true meaning of "going green." They will set up a solid foundation and fundamental framework of LEED for you. Each book in the LEED Exam Guides series covers every aspect of one or more specific LEED rating system(s) in plain and concise language and makes this information understandable to all people.

These pocket guides are small and easy to carry around. You can read them whenever you have a few extra minutes. They are indispensable books for all people—administrators; developers; contractors; architects; landscape architects; civil, mechanical, electrical, and plumbing engineers; interns; drafters; designers; and other design professionals.

Why is the LEED Exam Guides series needed?

A number of books are available that you can use to prepare for the LEED Exams:

1. *USGBC Reference Guides*. You need to select the correct version of the *Reference Guide* for your exam.

    The *USGBC Reference Guides* are comprehensive, but they give too much information. For example, *The LEED 2009 Reference Guide for Green Building Design and Construction (BD&C)* has about 700 oversized pages. Many of the calculations in the books are too detailed for the exam. They are also expensive (approximately $200 each, so most people may not buy them for their personal use, but instead, will seek to share an office copy).

    It is good to read a reference guide from cover to cover if you have the time. The problem is not too many people have time to read the whole reference guide. Even if you do read the whole guide, you may not remember the important issues to pass the LEED exam. You need to reread the material several times before you can remember much of it.

    Reading the reference guide from cover to cover without a guidebook is a difficult and inefficient way of preparing for the LEED AP Exam, because you do NOT know what USGBC and GBCI are looking for in the exam.

2. The USGBC workshops and related handouts are concise, but they do not cover extra credits (exemplary performance). The workshops are expensive, costing approximately $450 each.

3. Various books published by a third party are available on Amazon. However, most of them are not very helpful.

    There are many books on LEED, but not all are useful.

    LEED Exam Guides series will fill in the blanks and become a valuable, reliable source:

    a. They will give you more information for your money. Each of the books in the LEED Exam Guides series has more information than the related USGBC workshops.

    b. They are exam-oriented and more effective than the USGBC reference guides.

    c. They are better than most, if not all, of the other third-party books. They give you comprehensive study materials, sample questions and answers, mock exams and answers, and critical information on building LEED certification and going green. Other third-party books only give you a fraction of the information.

    d. They are comprehensive yet concise. They are small and easy to carry around. You can read them whenever you have a few extra minutes.

    e. They are great timesavers. I have highlighted the important information that you need to understand and MEMORIZE. I also make some acronyms and short sentences to help you easily remember the credit names.

It should take you about 1 or 2 weeks of full-time study to pass each of the LEED exams. I have met people who have spent 40 hours to study and passed the exams.

You can find sample texts and other information on the LEED Exam Guides series in customer discussion sections under each of my book's listing on Amazon.

What others are saying about *LEED GA Exam Guide* (Book 2, LEED Exam Guide series):

"Finally! A comprehensive study tool for LEED GA Prep!

"I took the 1-day Green LEED GA course and walked away with a power point binder printed in very small print—which was missing MUCH of the required information (although I didn't know it at the time). I studied my little heart out and took the test, only to fail it by 1 point. Turns out I did NOT study all the material I needed to in order to pass the test. I found this book, read it, marked it up, retook the test, and passed it with a 95%. Look, we all know the LEED GA exam is new and the resources for study are VERY limited. This one's the VERY best out there right now. I highly recommend it."
—ConsultantVA

"Complete overview for the LEED GA exam

"I studied this book for about 3 days and passed the exam ... if you are truly interested in learning about the LEED system and green building design, this is a great place to start."
—K.A. Evans

"A Wonderful Guide for the LEED GA Exam

"After deciding to take the LEED Green Associate exam, I started to look for the best possible study materials and resources. From what I thought would be a relatively easy task, it turned into a tedious endeavor. I realized that there are vast amounts of third-party guides and handbooks. Since the official sites offer little to no help, it became clear to me that my best chance to succeed and pass this exam would be to find the most comprehensive study guide that would not only teach me the topics, but would also give me a great background and understanding of what LEED actually is. Once I stumbled upon Mr. Chen's book, all my needs were answered. This is a great study guide that will give the reader the most complete view of the LEED exam and all that it entails.

"The book is written in an easy-to-understand language and brings up great examples, tying the material to the real world. The information is presented in a coherent and logical way, which optimizes the learning process and does not go into details that will not be needed for the LEED Green Associate Exam, as many other guides do. This book stays dead on topic and keeps the reader interested in the material.

"I highly recommend this book to anyone that is considering the LEED Green Associate Exam. I learned a great deal from this guide, and I am feeling very confident about my chances for passing my upcoming exam."
—Pavel Geystrin

"Easy to read, easy to understand

"I have read through the book once and found it to be the perfect study guide for me. The author does a great job of helping you get into the right frame of mind for the content of the exam. I had started by studying the Green Building Design and Construction reference guide for LEED projects produced by the USGBC. That was the wrong approach, simply too much information with very little retention. At 636 pages in textbook format, it would have been a daunting task to get through it. Gang Chen breaks down the points, helping to minimize the amount of information but maximizing the content I was able to absorb. I plan on going through the book a few more times, and I now believe I have the right information to pass the LEED Green Associate Exam."
—**Brian Hochstein**

"All in one—LEED GA prep material

"Since the LEED Green Associate exam is a newer addition by USGBC, there is not much information regarding study material for this exam. When I started looking around for material, I got really confused about what material I should buy. This LEED GA guide by Gang Chen is an answer to all my worries! It is a very precise book with lots of information, like how to approach the exam, what to study and what to skip, links to online material, and tips and tricks for passing the exam. It is like the 'one stop shop' for the LEED Green Associate Exam. I think this book can also be a good reference guide for green building professionals. A must-have!"
—**SwatiD**

"An ESSENTIAL LEED GA Exam Reference Guide

"This book is an invaluable tool in preparation for the LEED Green Associate (GA) Exam. As a practicing professional in the consulting realm, I found this book to be all-inclusive of the preparatory material needed for sitting the exam. The information provides clarity to the fundamental and advanced concepts of what LEED aims to achieve. A tremendous benefit is the connectivity of the concepts with real-world applications.

"The author, Gang Chen, provides a vast amount of knowledge in a very clear, concise, and logical media. For those that have not picked up a textbook in a while, it is very manageable to extract the needed information from this book. If you are taking the exam, do yourself a favor and purchase a copy of this great guide. Applicable fields: Civil Engineering, Architectural Design, MEP, and General Land Development."
—**Edwin L. Tamang**

Note: Other books in the **LEED Exam Guides series** are in the process of being produced. At least **One book will eventually be produced for each of the LEED exams.** The series include:

LEED GA EXAM GUIDE: *A Must-Have for the LEED Green Associate Exam: Comprehensive Study Materials, Sample Questions, Mock Exam, Green Building LEED Certification, and Sustainability* (3rd Large Format Edition), LEED Exam Guide series, ArchiteG.com (Published January 3, 2011)

LEED GA MOCK EXAMS: *Questions, Answers, and Explanations: A Must-Have for the LEED Green Associate Exam, Green Building LEED Certification, and Sustainability*, LEED Exam Guide series, ArchiteG.com (Published August 6, 2010)

LEED BD&C EXAM GUIDE: *A Must-Have for the LEED AP BD+C Exam: Comprehensive Study Materials, Sample Questions, Mock Exam, Green Building Design and Construction, LEED Certification, and Sustainability* (2nd Edition), LEED Exam Guide series, ArchiteG.com (Published December 26, 2011)

LEED BD&C MOCK EXAMS: *Questions, Answers, and Explanations: A Must-Have for the LEED AP BD+C Exam, Green Building LEED Certification, and Sustainability*, LEED Exam Guide series, ArchiteG.com (Published November 26, 2011)

LEED AP Exam Guide: *Study Materials, Sample Questions, Mock Exam, Building LEED Certification (LEED NC v2.2), and Going Green*, LEED Exam Guides series, LEEDSeries.com (Published on 9/23/2008).

LEED ID&C EXAM GUIDE: *A Must-Have for the LEED AP ID+C Exam: Comprehensive Study Materials, Sample Questions, Mock Exam, Green Interior Design and Construction, LEED Certification, and Sustainability*, LEED Exam Guide series, ArchiteG.com (Published March 8, 2010)

LEED O&M MOCK EXAMS: *Questions, Answers, and Explanations: A Must-Have for the LEED O&M Exam, Green Building LEED Certification, and Sustainability*, LEED Exam Guide series, ArchiteG.com (Published September 28, 2010)

LEED O&M EXAM GUIDE: *A Must-Have for the LEED AP O+M Exam: Comprehensive Study Materials, Sample Questions, Mock Exam, Green Building Operations and Maintenance, LEED Certification, and Sustainability (LEED v3.0)*, LEED Exam Guide series, ArchiteG.com

LEED HOMES EXAM GUIDE: *A Must-Have for the LEED AP Homes Exam: Comprehensive Study Materials, Sample Questions, Mock Exam, Green Building LEED Certification, and Sustainability*, LEED Exam Guide series, ArchiteG.com

LEED ND EXAM GUIDE: *A Must-Have for the LEED AP Neighborhood Development Exam: Comprehensive Study Materials, Sample Questions, Mock Exam, Green Building LEED Certification, and Sustainability*, LEED Exam Guide series, ArchiteG.com

How to order these books:
You can order the books listed above at:
http://www.GreenExamEducation.com

OR
http://www.ArchiteG.com

Building Construction

Project Management, Construction Administration, Drawings, Specs, Detailing Tips, Schedules, Checklists, and Secrets Others Don't Tell You (Architectural Practice Simplified, 2nd edition)

Learn the Tips, Become One of Those Who Know Building Construction and Architectural Practice, and Thrive!

For architectural practice and building design and construction industry, there are two kinds of people: those who know, and those who don't. The tips of building design and construction and project management have been undercover—until now.

Most of the existing books on building construction and architectural practice are too expensive, too complicated, and too long to be practical and helpful. This book simplifies the process to make it easier to understand and uncovers the tips of building design and construction and project management. It sets up a solid foundation and fundamental framework for this field. It covers every aspect of building construction and architectural practice in plain and concise language and introduces it to all people. Through practical case studies, it demonstrates the efficient and proper ways to handle various issues and problems in architectural practice and building design and construction industry.

It is for ordinary people and aspiring young architects as well as seasoned professionals in the construction industry. For ordinary people, it uncovers the tips of building construction; for aspiring architects, it works as a construction industry survival guide and a guidebook to shorten the process in mastering architectural practice and climbing up the professional ladder; for seasoned architects, it has many checklists to refresh their memory. It is an indispensable reference book for ordinary people, architectural students, interns, drafters, designers, seasoned architects, engineers, construction administrators, superintendents, construction managers, contractors, and developers.

You will learn:
1. How to develop your business and work with your client.
2. The entire process of building design and construction, including programming, entitlement, schematic design, design development, construction documents, bidding, and construction administration.
3. How to coordinate with governing agencies, including a county's health department and a city's planning, building, fire, public works departments, etc.
4. How to coordinate with your consultants, including soils, civil, structural, electrical, mechanical, plumbing engineers, landscape architects, etc.
5. How to create and use your own checklists to do quality control of your construction documents.
6. How to use various logs (i.e., RFI log, submittal log, field visit log, etc.) and lists (contact list, document control list, distribution list, etc.) to organize and simplify your work.
7. How to respond to RFI, issue CCDs, review change orders, submittals, etc.
8. How to make your architectural practice a profitable and successful business.

Planting Design Illustrated
A Must-Have for Landscape Architecture: A Holistic Garden Design Guide with Architectural and Horticultural Insight, and Ideas from Famous Gardens in Major Civilizations

One of the most significant books on landscaping!

This is one of the most comprehensive books on planting design. It fills in the blanks of the field and introduces poetry, painting, and symbolism into planting design. It covers in detail the two major systems of planting design: formal planting design and naturalistic planting design. It has numerous line drawings and photos to illustrate the planting design concepts and principles. Through in-depth discussions of historical precedents and practical case studies, it uncovers the fundamental design principles and concepts, as well as the underpinning philosophy for planting design. It is an indispensable reference book for landscape architecture students, designers, architects, urban planners, and ordinary garden lovers.

What Others Are Saying About *Planting Design Illustrated* ...

"I found this book to be absolutely fascinating. You will need to concentrate while reading it, but the effort will be well worth your time."
—**Bobbie Schwartz, former president of APLD (Association of Professional Landscape Designers) and author of *The Design Puzzle: Putting the Pieces Together.***

"This is a book that you have to read, and it is more than well worth your time. Gang Chen takes you well beyond what you will learn in other books about basic principles like color, texture, and mass."
—**Jane Berger, editor & publisher of gardendesignonline**

"As a longtime consumer of gardening books, I am impressed with Gang Chen's inclusion of new information on planting design theory for Chinese and Japanese gardens. Many gardening books discuss the beauty of Japanese gardens, and a few discuss the unique charms of Chinese gardens, but this one explains how Japanese and Chinese history, as well as geography and artistic traditions, bear on the development of each country's style. The material on traditional Western garden planting is thorough and inspiring, too. *Planting Design Illustrated* definitely rewards repeated reading and study. Any garden designer will read it with profit."
—**Jan Whitner, editor of the *Washington Park Arboretum Bulletin***

"Enhanced with an annotated bibliography and informative appendices, *Planting Design Illustrated* offers an especially "reader friendly" and practical guide that makes it a very strongly recommended addition to personal, professional, academic, and community library gardening & landscaping reference collection and supplemental reading list."
—**Midwest Book Review**

"Where to start? *Planting Design Illustrated* is, above all, fascinating and refreshing! Not something the lay reader encounters every day, the book presents an unlikely topic in an easily digestible, easy-to-follow way. It is superbly organized with a comprehensive table of contents, bibliography, and appendices. The writing, though expertly informative, maintains its accessibility throughout and is a joy to read. The detailed and beautiful illustrations expanding on the concepts presented were my favorite portion. One of the finest books I've encountered in this contest in the past 5 years."
—Writer's Digest 16th Annual International Self-Published Book Awards Judge's Commentary

"The work in my view has incredible application to planting design generally and a system approach to what is a very difficult subject to teach, at least in my experience. Also featured is a very beautiful philosophy of garden design principles bordering poetry. It's my strong conviction that this work needs to see the light of day by being published for the use of professionals, students & garden enthusiasts."
—Donald C. Brinkerhoff, FASLA, chairman and CEO of Lifescapes International, Inc.

Index

Made in the USA
San Bernardino, CA
14 April 2014